Introductio

London has thousands of pubs - don't count them, you'll get But what about their names? So many of them have such weird and wonderful names and what follows are the stories behind the names of over 650 of the capital's public houses.

We've omitted many obvious examples such as pubs named after monarchs. However, we've included popular pub names where there is an interesting story, such as Royal Oak or Red Lion.

We conducted a forensic search to pick out pub names and their meanings which required many hours of research, both flicking through books and websites but also on the ground investigation in the pubs themselves. We spoke to countless pub staff who have been very helpful; their patience and enthusiasm was really appreciated, not least given how difficult the last two years have been for the industry. We've endeavoured to give good coverage to all of London with pubs in every borough.

What we hope is evident is the huge variety in pub names across London and equally what inspired them. We're very proud to offer up tales of upstanding men and women. Plus, the more eccentric characters, customs, culture and aspects of London life.

We found that stories change with time and many are only oral tradition. We hope to document some of this aspect of London's social history and provide a snapshot in 2022. But we also take responsibility for any errors. If you have a suggestion, then please get in touch. James tweets @JamesPotts and Sam @InnsideTrack.

Anyway, that's enough from us. We'll let you dive in and enjoy learning more about the pubs in London and, in turn, a lot more about the city as a whole.

We hope you enjoy reading it as much as we enjoyed writing it... Cheers!

James Potts
Sam Cullen

Acknowledgements

We would like to thank those pub owners and staff who were kind enough to reply to our emails or put up with us asking where the name of their establishment comes from. Albert Jack, whose book on pub names was a vital source, and our friends at Londonist, whose website and book 'Londonist Drinks' was also key.

Other individuals kind enough to provide us with key snippets of information were: Chris Amies, Paul Cockle, Emma Dickinson and Mark Loughborough at Youngs, Simon Donoghue at Havering Local Studies Library, Andy Grant, Gerry Hahlo, Richard Holmes, Jane Jepchote, Alan Jepson, Andrew Sewell, Chris Shimwell at Greene King, Lisa Soverall at Southwark Archives, Georgina Wald at Fullers, Anthony Thomas from Antic, Robert Thomas from Remarkable Pubs and Poppy Wheldon.

There were several local history groups on Facebook that were also really helpful, in particular covering Bermondsey, Bow, Chessington, Sutton and Wimbledon.

We are also grateful to Jim Whiting at Capital History Publishing for his support and finally also Grace Bryant for proof reading the draft!

Aces & Eights *156-158 Fortess Road, Tufnell Park, NW5 2HP*
With the owners going for a saloon feel for this Tufnell Park pub, it is named after the Dead Man's Hand in poker, which is two aces and two eights. This is supposedly what folk hero Wild Bill Hickok had in his hand when he was murdered in 1876 in the Wild West of America.

Actress *90 Crystal Palace Road, SE22 9EY*
This name combines with a nearby pub owned by the same company to create a 19th century double entendre 'The Actress said to the Bishop'. For those not intimately acquainted with the old saying, it's not really an actress in the modern sense of the world, more someone taking part in the oldest profession in the world.

Admiral Codrington *17 Mossop Street, Chelsea, SW3 2LY*
Opened in 1830 and took its name from the Admiral who three years earlier had led the charge in the Battle of Navarino, where the British alongside the French and Russians had destroyed the Turkish and Egyptian fleet during the Greek war of independence.

Admiral Duncan *54 Old Compton St, Soho, W1D 4UD*
One of London's oldest LGBT venues, the Admiral Duncan honours the commander of the Royal Navy at the Battle of Camperdown in 1797. Adam Duncan was in charge when he defied convention by sailing his fleet directly at the opposing Dutch ships. This led to a famous victory for the Royal Navy and Duncan became a hero.

The Admiral Vernon *141 Broad Street, Dagenham, RM10 9HP*
Edward Vernon was a naval officer in the 18th Century. During the War of Jenkins' Ear, in 1739 he captured Porto Bello (now in Panama) from the Spanish with just six ships. In celebration, areas of London, Dublin and Edinburgh were named in its honour and Thomas Arne composed 'Rule Britannia'. A year later he instructed his sailors' rum to be diluted with water, leading to the concoction being labelled 'grog' after the fact Vernon was nicknamed Old Grog because he always wore a grogram coat.

Admiralty *66 Trafalgar Square, WC2N 5DS*
This takes inspiration from the nearby Admiralty Arch, built during Edward VII's reign to commemorate his mother Queen Victoria. The much older Admiralty House (built in 1788) is also located nearby.

Aeronaut *264 High Street, Acton, W3 9BH*
This pays homage to a daring Edwardian stuntman, George Lee Temple, who lived in Acton and was in 1913 the first man to successfully fly a plane upside down. An achievement worthy of recognition and how better to mark it than naming an establishment in his honour where if you have enough to drink you probably feel like you're flying upside down too.

Ailsa Tavern.

The Alpaca.

African Queen *315-317 Wellington Road South, TW4 5HL*
Takes its inspiration from the 1951 film of the same name, starring Humphrey Bogart and Katherine Hepburn, as several scenes were filmed at the nearby Isleworth studios. This film has also been cited as a reason behind parakeets establishing themselves in London, with the theory that some escaped during filming. This will not be the last time we see the parakeets in this book.

Ailsa Tavern *263 St Margarets Road, Twickenham TW1 1NJ*
Named after a former landlady of the pub, the story goes that she was engaged in a relationship with a local gentleman and they had to meet in secret due to the vast differences in their social status. Legend has it that there was an underground passage that linked the pub and our upstanding gent's house, and the two would have their secret rendezvous there. The romance part of the story could well be a myth, but an underground passage near the pub was discovered following excavations and local research in the 1960s.

Airman *Hanworth Road, Feltham, TW13 5AX*
The Airman harks back to the existence of the former London Air Park, which consisted of both a grass airfield and also significant facilities for the manufacture of aircraft in nearby Hanworth Park. The Air Park was at its height in the interwar period, coinciding with the pub opening for the first time in 1938. Given its proximity to the Air Park, it was in the line of fire during the Second World War and had one particular near miss when a German bomb landed in its carpark.

Alderman *52 Chippenham Road, RM3 8HX*
Thought to be named after the well-regarded local councillor Albert Dyer OBE who died three years before the pub opened in 1959.

Alexander Hay *3 Hay's Lane, Southwark SE1 2RA*
Situated in Hay's Galleria, this pub is named after the merchant who originally owned the site in the 1650s. In the 1840s, the lease changed hands and the new owner asked William Cubitt (brother of master builder Thomas Cubitt (see Thomas Cubitt) to turn it into a wharf for ships. One of the warehouses was leased to Frederick Horniman (see Horniman at Hay's). Hay's Wharf lasted until 1970 before the site was turned into Hay's Galleria in the 1980s, comprising shops, restaurants and this pub plus the Horniman at Hay's.

Alexander Pope *Cross Deep, Twickenham, TW1 4RB*
Alexander Pope was an 18th century poet who once had a grand country house and sprawling garden in this area; the pub is located on a portion of where his garden once stood. The original pub was destroyed by bombing in World War Two; the present building dates from 1959. Local controversy broke out in 2009 as the pub's name was changed from Pope's Grotto to simply the Alexander. The eventual compromise solution was the Alexander Pope. The Grotto it was named after is the only surviving element of Pope's once impressive riverside house and gardens. A committed band of local activists is seeking funds to open it up fully to the public all year round: at present it usually is accessible over heritage open days.

The Alfred Herring *316-322 Green Lanes, Palmers Green, N13 5TT*
Alfred Herring was a Second Lieutenant in the First World War who won a Victoria Cross for bravery in March 1918. After being surrounded with his men, he counter-attacked and held the advanced position for over eleven hours through the night. He kept encouraging his men to keep fighting which gave them the morale boost they needed. He was captured the next day and remained a POW for the rest of the conflict.

Alleyn's Head *Park Hall Road, SE21 8BW*
Edward Alleyn was a renowned actor in the Theatre during the Elizabethan era. In 1606 he purchased the Manor of Dulwich and 10 years later set up the College of God's Gift there, which is today's Dulwich College, as well as undertaking various other philanthropic activities. Due to his prominence in the late Tudor period, his character features in the 1998 film *Shakespeare in Love*, played by the American, Ben Affleck.

Allsop Arms *137 Gloucester Place, NW1 5AL*
This pub was named after the street it originally sat on, Allsop Terrace; it moved twice during the early 20th century, retaining its original name in a manner reminiscent of *Taggart* on ITV carrying on with the same title long after the lead character died.

The Alpaca *84-86 Essex Road, Islington N1 8LU*
Formerly the New Rose, renamed as the Alpaca just days before March 2020 saw the shutters come down for months. The incoming owners loved alpacas, not least because they'd had one of their first holidays together as a couple alpaca trekking in Norfolk. They also saw this furry animal as an ideal name for a community pub on account of them being such sociable creatures!

The Alwyne Castle *83 St Paul's Road, N1 2LY*
Spencer Joshua Alwyne Compton was the 2nd Marquess of Northampton who owned Canonbury Manor, which is near to where this pub can be found. He was cousin to Spencer Percival, the only Prime Minister to be assassinated, and he would later become President of the Royal Society. He even has a dinosaur named after him.

Anchor Tap *20A Horselydown Lane, SE1 2LN*
This Samuel Smith's pub is a stone's throw from Tower Bridge. Anchor in a pub name is usually there to attract sailors (who were notorious drinkers and could spend their wages in one night). The Anchor Tap was the taproom for the former Anchor Brewery, the building of which can still be seen between Tower Bridge and Butler's Wharf.

Ancient Foresters *282 Southwark Park Road, SE16 2HB*
This derives from the friendly society (a sort of community-led organisation providing banking, savings and other financial products) set up in 1834 called the Ancient Order of Foresters. The organisation still exists today, albeit with the shortened title of Foresters Friendly Society. Foresters is a more common variant of the pub name; those retaining the 'Ancient' prefix are rarer.

Angel in the Fields *37 Thayer Street, W1U 2QU*
The use of Angel as a pub name is a common one, indeed there are fourteen pubs that include angel in the title in London alone and like the Royal Oak an old pub lends its name to a tube station. The use of such a religious icon links back to travellers' hostels, given the protection they were supposedly offering. This particular one used to be The Angel and dates from the 1720s. The "in the fields" is an addition following a refurbishment in 2001 and hints back to the days when St Mary Le Bone was a small village just on the outskirts of London. Indeed, you can see the pub surround by fields on John Rocque's famous map of London from 1746.

Anglers *3 Broom Road, Teddington, TW11 9NR*
Owes its name to the fact this spot in the Thames has been particularly popular with fishermen over the years. Talking of fish, it's also right by where the famous Monty Python fish slapping dance skit was filmed, with Michael Palin ending up in Teddington lock after a very firm hit with a huge fish from John Cleese!

Angel in the Fields.

Antwerp Arms *168-170 Church Road, Tottenham, N17 8AS*
Originally known as the Hope and Anchor, this was renamed in 1861 after the Belgian city following a British brewery winning several prizes in a beer exhibition held there. The Antwerp almost vanished off the map for good in the early 2010s as the PubCo Enterprise planned to sell up to a property developer who'd convert it into flats. However, the community rallied behind the pub, setting up a co-operative to raise the money to buy it, which led to over 300 people investing as shareholders. One of these included Tottenham legend Gary Mabbutt who poured the first pint when it reopened in 2015, but there are countless others who deserve recognition.

The Arab Boy *289 Upper Richmond Road, Putney, SW15 6SP*
Named after Yussuf Sirrie, a servant of Henry Scarth, a local property developer who built the pub in 1849. Yussuf was later to become landlord of the pub after his former master left his entire estate to him in his will.

Aragon House *247 New Kings Road, SW6 4XG*
The pub reportedly stands on the site of the house Catherine of Aragon lived in after her first husband, Arthur, Prince of Wales, had died in 1506 and before she married Henry VIII in 1509.

The Archway Tavern *1 Navigator Square, Archway, Islington N19 3TD*
This pub is the focal point in the new Navigator Square and sits right next to the tube station of the same name. The name Archway comes from a short tunnel that existed prior to the Archway Road being built

as the world's first bypass in 1813. The name has stuck ever since and gives it name to the pub at the junction of Highgate Hill. The Kinks' 1971 album *Muswell Hillbillies* has a photo of the inside of the pub on its cover, despite not being in Muswell Hill.

Army & Navy *1-3 Matthias Road, Dalston, N16 8NN*
This could be named after either the two largest parts of the British Armed Forces or the co-operative society formed by officers in both organisations in 1871. It morphed into a department store group during the 20th century, including a particularly prominent building on Victoria Street just by the station. The brand finally vanished in 2005.

Artful Dodger *47 Royal Mint Street, E1 8LG*
Originally called the Crown and Seven Stars, the pub was renamed after the pickpocket from *Oliver Twist* in 1985 and where better for a thief than the one-time home of the Royal Mint. At the time this desire to tap into London and cockney heritage was all the rage, with names like Fagins, the Jolly Cockney and even Nice Little Earner popping up over the capital. While those have long since vanished, the Dodger still stands tall as a reminder of that era.

Artillery Arms *102 Bunhill Row, EC1Y 8ND*
Reflects the fact the pub is a mere stone's throw away from the Honourable Artillery Company's headquarters and grounds.

The Artillery Arms.

The Astronomer.

The Asparagus *1-13 Falcon Road, Battersea SW11 2PL*
You wouldn't know it looking at the area now but in the 19th century the trusty asparagus was synonymous with Battersea and the market fields that surrounded it - they were even packaged up as 'Battersea bundles'. Nowadays looking out of this pub, you're more likely to see a traffic jam on the busy York Road than a vegetable garden…

Assembly House *292-294 Kentish Town Road, NW5 2TG*
At the height of the reign of dastardly and squash-buckling highwaymen roaming the streets, travellers would gather where the pub stands today and then head up towards Hampstead Heath. The strength in numbers approach was hoped to give them a better chance against the Dick Turpins of this world.

Astronomer *125-129 Middlesex Street, E1 7JF*
The building the pub is housed in is called Astral House, hence Astronomer. It was previously called The Shooting Star so this is very much a continuation of the theme.

Atlantic Dog *389 Coldharbour Lane, Brixton SW9 8LQ*
This is a nod to both eras of this Brixton pub. It originally opened as The Atlantic, after the road it backs onto, but closed down in the early 1990s when the police revoked its licence following a series of disturbances there. In 1995 it reopened in a new guise as the Dogstar: one of the police's conditions to granting a new licence was the name was changed from The Atlantic. The new name was inspired by an article that Robert Harrison, the owner of the business, had been reading about how Sirius (the Dogstar) was the brightest star in the night sky, and instantly he knew it was the right fit for the pub. The link back to the pub's past came in 2020 when the name was amended to reinstate the Atlantic.

The Auld Shillelagh *105 Stoke Newington Church Street, N16 0UD*
An Irish pub in the middle of Stoke Newington, the Auld is the Gaelic for old and a shillelagh is a type of wooden club which can be used as a walking stick, or a weapon.

Avalon *16 Balham Hill, SW12 9EB*
When the new owners took over the former George pub in 2008, they wanted a distinctive name for their establishment. Looking for inspiration in historical texts by Shakespeare and Chaucer, it eventually came in the shape of a Led Zeppelin song, 'The Battle of Evermore', which referred to the 'angels of Avalon', Avalon being an island featured in Arthurian legend. Given Avalon's position in British folklore, they felt it was a perfect fit for the pub.

Axe *18 Northwold Road, Stoke Newington, N16 7HR*
A relatively rare example of a pub name inspired by archaeological finds, as various stone age flint axe heads were discovered near to the site of the pub and several other locations in Stoke Newington.

Bag O'Nails *6 Buckingham Palace Road, SW1W 0PP*
A bag of nails was the symbol used by ironmongers to advertise their trade.

Balham Bowls Club *7-9 Ramsden Road, Balham, SW12 8QX*
As the name suggests, this pub was once a bowls club house. Locally it's often known as the BBC, although you don't need to pay the licence fee to enter.

Ballot Box *Horsenden Lane North, Greenford, UB6 7QL*
You're probably used to casting your vote in elections at your local church, primary school or community hall, but how about the pub? That's exactly what residents of the canal boats in Greenford did in the 19th century. This is not the original pub where voting took place: the name was carried over from a pub which was located on the other side of the hill. Curiously enough it now has a Wacky Warehouse as part of the pub, this being a soft play centre for kids as opposed to a satirical statement on modern British democracy.

Bank of Friendship *226 Blackstock Road, Islington N5 1EA*
One of London's lost rivers is responsible for this name. The Hackney Brook runs nearby before it empties into the River Lea. Supposedly the residents of Highbury would wave to their neighbours in Stoke Newington over the other side of the river, giving its name to the pub.

Barley Mow *Various - Five Pubs in London*
Mow in this sense refers to a stack of barley which, of course, is a key ingredient of beer. Barley Mow is also the name of a folk song that would have been sung in pubs across the UK.

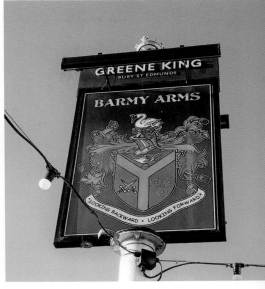

Bank of Friendship.

Barmy Arms.

The Barrel Vault *Unit 23, St Pancras International Station, N1C 4QP*
A new addition to the roster of Wetherspoons in London, the Barrel
Vault hints at St. Pancras's history. The large concourse underneath
the platforms where the pub can be found used to be the vaults where
the goods coming into London were stored before being distributed.
These included large barrels of beer from the East Midlands. The space
was ingeniously brought to life when the station was redeveloped in
the 2000s.

Barmy Arms *The Embankment, Twickenham, TW1 3DU*
Originally the Queens Head, this gained its present name in the late
1970s in honour of a previous owner who put up his Christmas tree
upside-down. Whether the tree remained upside-down throughout the
month of December is sadly not documented.

The Barking Dog *61 Station Parade, Barking, IG11 8TU*
Although this might be hard to believe, the name of this pub does not
have anything to do with dogs but instead it's related to fishing. This is
because Barking's main industry for centuries was fishing, where the
Barking locals would fish off Doggerbank in the North Sea.

Barrow Boy & Banker *6-8 Borough High Street, SE1 9QQ*
One of many bank-to-pub conversions across London, the name here
reflects both the building's previous origins as well as its proximity to
Borough Market. These days the market is decidedly more upmarket,
so you'll be more likely to encounter artisan cheesemongers there than
any barrow boys.

Baxter's Court *282-284 Mare Street, E8 1HE*
Hackney's Wetherspoons sits on the site of an old alleyway called
Baxter's Court. The pub pays homage to this.

Bear & Staff *10-12 Bear Street, WC2H 7AX*
Also called the Bear and Ragged Staff elsewhere, this name has a
connection to Richard Neville who was the Earl of Warwick in the
1400s. He is known as the Kingmaker in Shakespeare's Henry VI and
the symbol comes from his coat of arms which depicts a bear and a staff
which is refenced in the play and features on the flag of Warwickshire.
The legend as to where this comes from says that the first Earl strangled
a bear and the second killed a giant using a ragged staff hence why they
are both featured.

The Beaten Docket *50-56, Cricklewood Broadway, NW2 3ET*
A Beaten Docket is a losing betting slip, usually associated with
horseracing which was a feature in Cricklewood in the late 19th
century. Wetherspoons thought this would make a good name for a
pub, and they'd be right.

The Beaufort *2 Heritage Avenue, Grahame Park Way, NW9 5AA*
The Beaufort is the name of a dukedom after a castle in Champagne,
France. They are descended from John of Gaunt (See The Red Lion)
who owned the castle at the time. Charles II was the first to bestow this
upon the first duke in 1682. It remains the only dukedom to be named
after somewhere else not in the UK.

The Barrowboy &
Banker.

The Bear & Staff.

Beau Brummell *3 Norris Street, St James, SW1Y 4RF*
A new entrant to the London pub scene in 2021, the owners decided that the stylish dandy and one-time close friend of the Prince Regent, who lived just up the road from their pub in Mayfair, was just the person to name their pub after. The man himself once recommended that boots should be polished with champagne - we're sure the pub would kindly oblige if anyone wanted to try the true Brummell experience...

The Beehive *Various - Eleven pubs in London*
There are 11 pubs with this name buzzing around London. Usually this name is thought to derive from the fact it is a concept that lends itself very well to being represented on a pub sign. Brentford's Beehive has gone a step further and has a golden, beehive shaped dome, at the top of the pub. That entry in the hive is named after the Beehive Brewery, which was operated on the Brentford area from the mid-19th century until being acquired by Fullers in 1907.

The Betjeman Arms *53 Euston Road, N1C 4QL*
Sir John Betjeman was a famous poet during the 20th Century, serving as Poet Laureate between 1972 and 1984. Following the destruction of the arch at nearby Euston (see The Doric Arch), Betjeman campaigned to save St Pancras station which was firmly in the bulldozer's sights. This campaign was ultimately successful and the station has since been refurbished to its former glory. The pub that bears his name now sits on the concourse which he fought so hard to save.

Betsey Trotwood *56 Farringdon Road, EC1R 3BL*
Betsey was a character in Charles Dickens classic *David Copperfield*. Trotwood was the title character's great aunt and who has a singularly low opinion of the male species after having a bad experience of marriage. She does come good for Copperfield in the end, though. Why was the name chosen for this pub? Clerkenwell is deep in Dickens territory and the pub is only a few picked pockets away from Pear Tree Court, which is thought to be the inspiration for where Oliver Twist saw the Artful Dodger hard at work.

Betsey Trotwood.

Beulah Spa *41 Beulah Hill, Norwood, SE19 3DS*
Harks back to the existence of a spa on Beulah Hill which was a big hit
in the early Victorian period, with the Queen herself even visiting on
four occasions. During the attraction's peak, there were three licensed
establishments vying for trade within a short proximity of each other.
While the current building here opened in 1939, it is considered the
successor to the Beulah Tap Room, the venue for the staff and coachman
rather than those able to afford to use the spa.

Bill Murray *39 Queen's Head Street, Angel, N1 8NQ*
Opened in 2017 to provide a permanent base for the Angel Comedy
Club, the Bill Murray is also a fully functioning pub open seven days
a week. The Club were able to realise their dream of acquiring the
venue thanks to an internet kickstarter campaign. Officially the pub is
said to be named after William Murray, Charles I's childhood whipping
boy - which quite literally meant he received the punishments Charles
should have received but was exempt because of his royal status.
Given the comedy connections, we think it would be hard for the real
inspiration for the pub name to be lost in translation.

Bill Murray.

Bill Nicholson *102 Northumberland Park, Tottenham, N17 0TS*
Nicholson was the most successful manager in Tottenham Hotspur's history, winning eight trophies in his 16-year tenure including, most significantly, the first league and cup double of the 20th century in 1960-61, so it seems only right he gives his name to this pub a stone's throw from their shiny new stadium. As long-suffering Spurs fans need no reminding of, he remains the last manager to win them the league.

Birds *692 High Road, Leytonstone, E11 3AA*
A tribute to arguably the most famous man to hail from Leytonstone, Sir Alfred Hitchcock (some might say he has now been eclipsed by David Beckham), this pub has been named after his 1963 film where our feathered friends turn on the population of a town in California.

Bird Cage *80 Columbia Road, E2 7QB*
Derived from the silk-weaving Huguenot immigrants who settled in this part of town in the late 17th century and brought with them their tradition of keeping caged songbirds.

Bird's Nest *32 Deptford Church Street, SE8 4RZ*
During its long history as the Oxford Arms, the pub was a thriving live music venue, the roll of honour including local boys Squeeze and not quite so local Dire Straits. The name changed in the early 2000s and is thought to refer to the pub's location next to Deptford Creek, which was itself a hotbed of shipbuilding in the 18th century. Technically, the observation point at the top of a ship is a crow's nest, but what is a crow if it is not a bird?

Bird and Barrel *100 Barnehurst Road, DA7 6HG*
Micropub of the Bexley Brewery, the name derives from their logo — the green Ring Necked Parakeet, and the fact their beers are available on tap here, by the barrel! There are various stories and urban myths about how parakeets were first unleashed on London; we covered one earlier at the African Queen but another involves Jimi Hendrix letting two free in Soho in 1967 and then them multiplying from there. Sadly, a trio of London Universities debunked these myths in 2019 and stated there were first sightings way back in the 1860s. Either way, they are a common sight in many parts of the capital, with 32,000 being recorded in their most recent 'roost count' (sadly there are no pubs of that name in London!) back in 2012.

Birkbeck Tavern *45 Langthorne Road, Leyton, E11 4HR*
The last remaining 'Birkbeck' pub in London, in the 19th century there was a handful of taverns all bearing the Birkbeck name on estates built during the mid-Victorian period by the Birkbeck bank. The bank was closely interlinked with the London Mechanics Institute, the precursor to today's Birkbeck College in Bloomsbury, and initially shared the same offices. We will come across a former Birkbeck Tavern later.

Bishop *Off Bishop's Hall, Thames Street, Kingston, KT1 1PY*
Previously known as The Bishop Out Of Residence, this referred to the fact that the Bishop Of Winchester at the end of the 14th century had a palace (the standard name for a bishop's house) built right by where the pub stands. However, he rarely made use of an apartment which today's estate agents would label as 'a fantastic riverside setting' and so was usually out of that residence, hence the pub's old name. While the shorter name is undeniably snappier, our romantic side can't help but wish for the 'Out of Residence' to return to this residence in the future...

Bishop's Finger *9-10 West Smithfield, Smithfield Street, Farringdon, EC1A 9JR*
The bishop in question here is Thomas Becket who was murdered by followers of King Henry II in 1170 after the king famously spoke aloud: "will no one rid me of this turbulent priest?", having fallen out with Becket. His men took that literally and murdered the bishop, sparking outrage from the church. The fingers were found on signposts directing followers to his shrine, which existed until Henry VIII dissolved the monasteries in the 1530s.

The Black Prince *6 Black Prince Road, Kennginton, SE11 6HS*
Edward, the Black Prince, was the son of King Edward III, and his heir. A key military commander during the Hundred Years War, he was the brother of John of Gaunt (See The Red Lion). He would die of dysentery aged 45. As such, when his father died the crown passed to his son, who became Richard II. This pub was famously used in the movie *Kingsman: The Secret Service*.

The Blackfriar.

The Blackfriar *174 Queen Victoria Street, Blackfriars, EC4V 4EG*
The pub, and indeed the mainline train station nearby, get their name as there once was an order of Dominican Monks who built a monastery in the area. Their robes were black, hence the name. The monastery was shut down during the Dissolution of the Monasteries in 1538 but the name has stuck to the area ever since.

The Black Horse *Various, 13 Pubs in London*
There is a baker's dozen of Black Horse pubs across London. The name is thought to date back to the 14th century and its roots are thought to be as simple as the fact it lend itself easily to being depicted on an inn sign. A Black Horse also has the the distinction of being one of the three animals featured in the name of a Tube station, at Blackhorse Road, the others being a Canary (Wharf) and a Raven(scourt Park).

The Blind Beggar *337 Whitechapel Road, Shadwell, E1 1BU*
Famous not only for being the site where the Salvation Army finds its origins following a sermon from William Booth in 1865 but also a little over 100 years later when Ronnie Kray shot George Cornell in the bar. Legend has it that the name comes from 'The Ballad of Bethnal Green', an Elizabethan poem which tells the tale of Bessie whose father is a blind beggar. As such, none of the knights will marry her, until a humble one agrees. At this point the beggar reveals himself to be the Earl of Leicester and ensures the couple have a fortune for their future. A happy ending for once!

The Black Horse.

Bloated Mallard *147 High Street, Hampton, TW12 1NJ*
Originally the name of a different pub located just over a mile away by Teddington railway station, there are two reasons behind the name. The Mallard element was inspired by the railways, being the name of a 1938 steam locomotive which holds the world speed record for this type of vehicle at 126mph, which was achieved on 3 July 1938. The Bloated prefix was added to parody Heston Blumenthal's high-end restaurant, 'The Fat Duck'. Hampton pub was originally meant to be the second of the Mallards, however, the original went under before this one opened. Thankfully for our purposes they chose to keep the name regardless.

Bluecoats *614 High Road, Tottenham, N17 9TA*
This offers us a history lesson back to the original use of this building,
which first opened in 1835 as The Bluecoats School for Girls.

The Blue Posts.

The Bountiful Cow.

Blue Posts *Various - Five Pubs in London*
There are a number of Blue Posts in central London. Blue posts used
to historically be a form of taxi rank where men carrying sedan chairs
could be hired to take the wealthy home after a night out (see also Two
Chairmen).

The Bobbin *1-3 Lilleshall Road, Clapham. SW4 0LN*
Originally called Tim Bobbin, Tim was dropped in the early 2000s.
Mr Bobbin was actually the pseudonym of John Collier, an English
satirical poet and caricaturist hailing from Lancashire and styling
himself as that region's Hogarth. Now Hogarth's links to London and
drinking is well documented, but how did a pub in Clapham end
up being named after his Lancastrian equivalent? A 2001 magazine
article revealed an intriguing theory. Collier's first biographer, Richard
Towney, had relatives living in Turret Place in Clapham, and would
frequently visit and bring examples of Bobbin's work. Their house was
a stone's throw away from Lilleshall Road. The article speculated that
perhaps someone connected with the household ended up buying the
pub and naming it after Bobbin, having been so taken with his artwork.

The Boleyn Tavern *1 Barking Road, West Ham, E6 1PW*
The pub, and indeed West Ham's former stadium, get their name from
the fact that Anne Boleyn at least visited, at most owned, the nearby
Green Street House which locally became known as Boleyn Castle. The
pub is obviously frequented by West Ham fans on match days and was
visited by Gandhi in 1931. He had a cream soda apparently!

Bolt Hole *12 Falconwood Road, Welling, DA16 2PL*
This reflects the ethos behind the micropub, which clearly states on its website that it's 'a place a person can escape and hide'.

The Botwell Inn *25-29 Coldharbour Lane, Hayes UB3 3EB*
Botwell was a hamlet that used to sit on the site of Hayes town centre, giving the name to this Wetherspoons.

The Boogaloo *312 Archway Road, Highgate, N6 5AT*
A Boogaloo is a type of funk dance that originated in the USA in the 1960s, with an 'electric' variant becoming popular in the 1980s. The name also hints at the fact the pub hosts gigs, has a famous jukebox and even its own radio station, showing its worthy musical credentials. The pub began life as a Birkbeck Tavern, this was for the same reason as the pub of that name which is still trading in Leyton. The Birkbeck name here is preserved in tiling by the pub's entrance.

Bountiful Cow *51 Eagle Street, WC1R 4AP*
Opened in 2004 by the fantastically named Roxy Beaujolais. At the time, she proclaimed it to be 'devoted to beef', and the cow theme is continued throughout the interior with posters from famous and not so famous cowboy films, as well as plenty of assorted bovine iconography. It has since lent its name to a media agency based a short distance away which started up in 2016, as the owners were big fans of the pub! The company was a spin-off from their earlier business, named The7stars after another nearby pub which was also owned by Roxy. (See The Seven Stars).

Bourbon *387 Roman Road, E3 5QR*
A recent entry having opened in Autumn 2021, this name reflects the theme of the pub as a US dive bar, complete with American beers, exposed brickwork and pictures of soul singers on the wall, and, of course, plenty of varieties of bourbon.

The Bow Bells *116 Bow Road, Bow, E3 3AA*
This quintessential East End pub is named after the famous bells in St Mary-le-Bow church in the City of London. The bells were used to order a curfew on residents of the city before the church was burnt down during the Great Fire and was subsequently rebuilt by Christopher Wren. They could also be heard beyond the city, which made Dick Whittington turn around and come back to London (See The Whittington Stone). You were also defined to be Cockney if you were born within earshot of the bells.

Brave New World *22-26 Berrylands Road, KT5 8RA*
Despite our best efforts, we weren't able to find out the story behind this Surbiton watering hole. The most obvious connection would be a homage to Aldous Huxley's literary classic. However, we've kept

this in the book for the sole fact it's directly opposite the office of the current leader of the Liberal Democrats, Sir Ed Davey, meaning if his staff ever have an idle moment and stare out of the window, they see a sign saying BRAVE NEW WORLD looking back at them, inspiration enough to keep the yellow flame alive...

Brass Monkey *250 Vauxhall Bridge Road, SW1V 1AU*
Looked at in a certain light, this pub could resemble a room on a ship, complete with its curved frontage windows. It is that theme that an earlier owner of the pub ran with, choosing to call it after what is alleged to be the origin of the term 'brass monkey', a kind of storage rack for cannonballs on a ship. While this is widely now thought to be an urban myth, monkey was a term for a variety of guns and cannons in the 17th century, so it is understandable how the story grew up. True or not, it certainly makes for an intriguing pub name.

Brave Sir Robin *29 Crouch Hill, Finsbury Park, N4 4AP*
An ironic name here. Brave Sir Robin is from the 1975 film *Monty Python and the Holy Grail* who runs away at the first sign of danger. In the film he was played by Eric Idle.

Bread & Roses *68 Clapham Manor Street, SW4 6DZ*
Owned by the Workers Beer Company, the Bread & Roses takes its name from a speech given by suffragette Helen Todd in the USA in 1910. The phrase was then turned into a poem which was later put to music, becoming an anthem for left wingers ever since.

Brendan the Navigator.

The Bricklayers Arms.

Brendan the Navigator *90 Highgate Hill, Highgate, N19 5NQ*
A new Irish gastropub, Brendan the Navigator was a sixth century priest who set out on a quest to find the Garden of Eden. He is the patron saint of travellers and sailors.

Bricklayers Arms *Various - Nine pubs in London*
Several pubs remember the tradition of bricklaying. Given the need for bricks to build London, bricklayers would have been kept busy and would have needed a place to drink. As with many "arms" pubs, the name could reflect the occupation of the landlord.

Britannia *1 Allen Street, W8 6UX*
When a pub first opened on this street in 1834, it was the tap room of the Britannia brewery occupying the wider site. The brewery was demolished in the 1920s after being acquired by Youngs and replaced by the Allen Mansions block of apartments but lives on in the memory via this pub name. It seems very strange to think a brewery was once based on a well-heeled residential street by Kensington High Street.

Broadcaster *101 Wood Lane, White City, W12 7FW*
A nod to the past use of this site, this pub is based in a new building that has sprung up as part of the wider redevelopment of the old BBC Television Centre. While the majority of the site has been converted into housing, three studios do still remain in BBC hands and enough of the old building is still intact so you can still picture Alan Partridge running out of the building shouting 'I've got cheese!' after his disastrous meeting with Tony Hayers.

Brockley Barge *184 Brockley Road, Brockley, SE4 2RR*
A barge often conjures images of boats on water, yet there is no large body of water near Brockley. So why the name of this pub? Well, the railway line to Croydon was formerly a canal. Opening in 1809, the Croydon Canal was not commercially successful and closed in 1836, to be replaced by the railway.

Brockley Jack *408 Brockley Road, Brockley, SE4 2DH*
This pub is reputedly named after Jack Cade, who led a rebellion in 1450. He tried to keep his men in check but upon reaching London they began to loot and were pushed back by the citizens of London to London Bridge. Cade was later caught by the Sheriff of Kent and was fatally wounded before he could stand trial.

Broken Drum *308 Westwood Lane, Sidcup, DA15 9PT*
This Sidcup micropub opened its doors in 2015, taking its name from a pub in the first books of Terry Pratchett's legendary Discworld series. Thankfully, the south-east London version has fared better than its literary inspiration. Pratchett's version was destroyed in the great fire 'Ankh-Morpork', subsequently reopening as 'The Mended Drum'.

Brouge *214 Hampton Road, Twickenham, TW2 5NJ*
This distinctive looking word was chosen specifically for an English and Belgian gastropub, to reflect the fusion of both cuisines, Brouge being adopted as an anglicisation of Bruges. It's been Brouge here since 2012; for a spell Brouge was just the name of the restaurant and the pub retained its then name of the 'Old Goat', which was what the former owner's father used to call his mother. Answers on a postcard please for what the Old Goat referred to him as...

Brunel *47 Swan Road, Rotherhithe, SE16 4JN*
Named after Marc Isambard Brunel, the man behind the construction of the nearby Thames Tunnel, the first tunnel under the river, and father of Isambard Kingdom, the most famous of the Brunels.

Buff *Pinewood Drive, Orpington, BR6 9NL*
This one's a zinger. Deriving from a breed of chicken named the 'Orpington' originally bred in the colour buff (light brownish yellow to those without your dulux chicken colour charts) by a William Cook who hailed from these parts. The chicken comes in four other colours but the Buff stands above the rest as the most common of the 'Orpingtons'.

Bull & Gate *389 Kentish Town Road, NW5 2TJ*
A case of mispronunciation here, the pub was originally called The Boulogne Gate but over the years the name Bull & Gate stuck until it was made official. The pub pays homage to its origins, though, by having a cocktail bar upstairs called the Boulogne Bar.

Bull & Last *168 Highgate Road, NW5 1QS*
The 'Last' part of this name is what sparks curiosity. It apparently comes from the fact the pub was the last one as you headed up the Great North Road out of London. A coachman would shout "last" to indicate that this was the last one before leaving the capital. This pub is not to be confused with other pubs near Hampstead Heath, such as the Bull & Gate and the Bull & Bush.

Butcher's Hook and Cleaver *60-63 Smithfield, EC1A 9DY*
A suitably meaty name right by Smithfield Market, half of this pub building was a meat supply shop. The other half was a bank. If Fullers ever want to rename this pub, the Butcher's Bank Balance would be a great alliterative title.

Butterchurn *Erskine Road, Sutton, SM1 3AS*
Multiple sources were churned through in an attempt to uncover the meaning behind this pub name, the only Butterchurn in the UK, no less! In the absence of any formal records that could help, the best explanation we've heard is that there was once a dairy located on the parade of shops close to the pub.

The Cabbage Patch.

Cabbage Patch *67 London Road, Twickenham, TW1 3SZ*
Rugby aficionados will know this story but for the uninitiated, this vegetable based name comes from the fact that the land that Twickenham Rugby Stadium was built on was previously used as allotments with cabbages being a staple product, so Cabbage Patch became one of the ground's affectionate nicknames and, in turn, the name of this pub too.

The California *267 Brighton Road, Belmont SM2 5SU*
There are two competing stories as to why this pub is called the California, both involve a Mr J Gibbon and gold but beyond that they diverge significantly. One contends that Gibbon discovered some gold coins while working in a nearby lime mine, used these to get to California, made his fortune in the gold rush then returned home and opened this pub.

    The other story also states Gibbons made his fortune in the gold-rush but this time was only in California after fleeing his trial for poaching a pheasant off a country estate in nearby Cheam! Regardless of what view you take of Gibbons, it is certainly true that the pub initially ended up lending its name to the area immediately around it. This led to the railway station initially being called California, too, when it opened in 1865. It changed to its present name of Belmont in 1875 and sadly removed the novelty of saying you're hopping on a train from Clapham Junction to California…

Camel and Artichoke *121 Lower Marsh, SE1 7AE*
Originally known as the Artichoke on account of it being located on one of the main routes that artichokes were transported through London. Around the turn of the millennium it had a brief spell as the Elusive Camel before new owners came in and decided to blend the old with the new. There were two other Elusive Camels in London at this point, one near Victoria Station and another by London Bridge, so not sure it's fair to say they were that elusive…

Candlemaker *136 Battersea High Street, SW11 3JR*
Previously known as the Greyhound, this pub now references the former
Prices Candle Factory located a few minutes along the road. Elements
of the factory remain as the 'Candle Factory Apartments'. The sight of
the derelict factory awaiting demolition and redevelopment back in
2003 inspired Paul Talling to start his 'Derelict London' website which
has spawned two excellent books covering a whole host of disused
buildings across the capital.

The Capitol *11-21 London Road, Forest Hill, SE23 3TW*
Another Wetherspoons which takes its name after the cinema which
used to occupy the building until the 1970s. It then became a bingo
hall during the 80s and 90s, before becoming a pub. (See The Coronet).

Captain Cook *205 Dawes Road, SW6 7QY*
Once known as the Wilton Arms, the pub was derelict by the mid-
2010s until being taken over by a duo hailing from New Zealand, Chris
Wilkie and Karl Tisch, who revived and renamed it in honour of the
English navigator who was the first European to discover and map
New Zealand, reaching its shores in 1769.

Captain Kidd *108 Wapping High Street, E1W 2NE*
Another of Wapping's riverside nautical pubs, Captain Kidd is named
in honour of William Kidd, the 17th Century Scottish pirate who was
tried and executed in 1701. His legend grew from there and he inspired
many writers, including Robert Louis Stevenson when he was writing
Treasure Island.

The Case is Altered *Eastcote High Road, HA5 2EQ*
and Old Redding, Harrow, HA3 6SE
Several pubs across England have this most peculiar of names, each
with their own theory as to how it came about. However, the most
likely source comes from Edmund Plowden who was a catholic lawyer
living under Elizabeth I's protestant reign. His skills as a lawyer spared
his life after impressing the queen by defending his faith. His other
moment of fame came when defending a catholic who had been
accused of taking mass. However, as the trial went on, it became clear
that it was conducted by a layman and not a priest, and can therefore
not constitute an official mass. He is said to have said: "No priest, no
mass! The case is altered!".

The Catcher in the Rye *317 Regents Park Road, N3 1DP*
Named, of course, after J D Salinger's novel of 1951. The pub opened
in the mid-1990s but sadly we were unable to find out why this pub
is named after the novel: even those connected with the original
landlords were unable to shed any light on the matter.

The Chandos.

Cat and Mutton *76 Broadway Market, Hackney, E8 4QJ*
Named after a bridge over the Regents Canal a short distance from the pub. That structure takes its name from the fact coal barges passing under the bridge were referred to as cats, and the mutton aspect for the various people trooping through the area with their livestock to take to sell at the London markets. Previous names for the pub have all been variations on the same theme, such as Cat and Shoulder of Mutton or the Leg of Mutton and Cat.

The Cat's Back *86-88 Point Pleasant, Wandsworth SW18 1PP*
A name as intriguing and puzzling as our feline friends can be. The story told to us by the current landlord which was recounted to him when he purchased the pub is that the pub's cat in the inter-war period disappeared and then returned years later. People speculated that the intrepid feline had stowed away on a boat and travelled to exotic lands before tiring of a life on the high seas and deciding to return to the creature comforts of Wandsworth.

Cavern *100 Coombe Lane, SW20 0AY*
This pub is themed as a 1960s rock bar and so pays tribute to arguably the most famous venue of that era, synonymous with the Beatles and Liverpool, and, now, Raynes Park...

Centre Page *29-33 Knightrider Street, EC4V 5BH*
Formerly part of the long since defunct 'Front Page Pub' group, this is now one of only three of the chain's old pubs that has retained its page themed name under different ownership. The name here refers to the former dominance of the newspaper industry on Fleet Street. You'll find the other two later on in this book...

The Chandos *29 St Martin's Lane, Westminster, WC2N 4ER*
This Sam Smith's pub is named after the Dukes of Chandos who were instrumental in developing this area of London during the early 18th century. Their family were first awarded a peerage way back in the 1330s, at that time Baron Chandos. This particular lineage came to an end in 1511, before being revived in 1554.

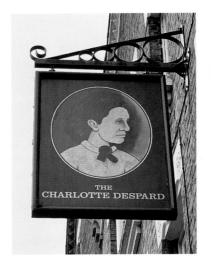

The Charlotte Despard

The Churchill Arms.

Charles Holden *198 High Street, Colliers Wood, SW19 2BH*
This takes its name from the man who designed Colliers Wood tube station right opposite the pub. Holden was involved in the design of over 40 other stations across the capital and his architectural styles and use of Portland Stone became synonymous with the expansion of the Underground network into the suburbs during the 1920s and 30s. His legacy is also felt on a much more imposing scale on the London skyline in the shape of both 55 Broadway (former headquarters of TfL and predecessor bodies) and, most strikingly, Senate House in Bloomsbury, often used in dystopian dramas as the HQ for totalitarian regimes. In reality, the University of London and its library are more benign.

The Charlotte Despard *17-19 Archway Road, Archway, N19 3TX*
Charlotte Despard (1844-1939) was a suffragist and activist who campaigned on numerous issues throughout her life in both Ireland and Britain. She helped found the Women's Freedom League in 1907 and was a member of the Labour Party, standing for Parliament after WWI. In her later years she joined the Communist Party.

Chelsea Pensioner *358 Fulham Road, SW10 9UU*
While this pub is located far closer to Stamford Bridge than the Royal Hospital Chelsea, the name is a nod to former members of the British Army who are residents at the latter, which is both a retirement and nursing home for ex service personnel. (see also The Pensioner)

Churchill Arms *119 Kensington Church Street, Notting Hill, W8 7LN*
One of the most recognisable pubs in London thanks to its year-round floral displays, which have received awards in the Chelsea Flower Show, and for its extravagant Christmas tree lights during the festive

season. It was called the Marlborough Arms up until after World War Two, when it was renamed the Churchill Arms. Technically, the Churchill in question it's named after is Sir Winston's grandfather, John Spencer-Churchill, 7th Duke of Marlborough, who used to drink here in the 19th century. You now can't move for pictures of Britain's wartime PM inside this west London pub.

The Clachan *34 Kingly Street, Carnaby, W1B 5QH*
Clachan is the Gaelic word for a meeting place. Sounds like a good name for a pub!

City Barge *27 Strand-on-the-Green, W4 3PH*
There has reputedly been a pub on this site since 1484, initially called the Navigation Arms then renamed to Bohemia Head in 1618, in honour of Frederick V who was a hero in Britain for having just conquered Bohemia and for being married to James I's daughter, Elizabeth. It acquired its present name in 1788 due to the fact the Lord Mayor of London's ceremonial barge was kept on the other side of the river during the winter months.

Clock House *196a Peckham Rye, SE22 9QA*
A nod to a part of the internal fabric of this building which has survived its evolution from shop to off licence and finally pub, there is a large Victorian clock which dominates the interior of this Peckham public house.

The Coal Hole *91-92 Strand, WC2R 0DW*
The Coal Hole sits on the Strand next to The Savoy. Its name apparently comes from an early 18th Century landlord who would host 'song and supper nights' and was known as the singing collier. However, the pub is also where the coal for the hotel used to be dropped off, so this pub could have a second meaning!

The Coal Hole.

The Cocoanut.

Cockpit *7 St Andrew's Hill, EC4V 5BY*
The pub building was originally used for hosting cock fights until the practice was outlawed in 1835, the cockpit being the name for the arena where the male chickens fought it out. The gallery where spectators watched proceedings above the main cockpit area can still be seen today.

Cocoanut *16 Mill Street, Kingston, KT1 2RF*
This is definitely not a typo - cocoanut being an earlier spelling of Coconut and a reflection of the fact there was once a bustling trade in coconut fibre in a factory at the bottom of the same street the pub is on, backing onto the Hogsmill River. The factory was a major employer in the area during the mid-19th century, when the pub was renamed from the Joiners Arms in 1858. There are also rumours the first coconut shy took place in Kingston, too...

The Colonel Fawcett *1 Randolph Street, NW1 0SS*
Formerly called The Camden Arms, this pub remembers one of the final duels that took place in the UK in 1843. Fawcett was challenged to a duel by a fellow army officer, Alexander Munro, after Munro insulted Fawcett's wife. Fawcett was found with a chest wound; he made his way to the Camden Arms where he died two days later. Munro fled the country but returned to face justice where he was sentenced to death but managed to get that down to just a year in prison. The pub now honours Fawcett.

Colton Arms *187 Greyhound Road, W14 9SD*
The Colton in the name here comes from its first licensee, George Colton Moor, who first opened the pub here in 1849 and made clay-pipes before he started pulling pints in this corner of West Kensington.

Conductor *1 Fleet Place, EC4M 7RA*
You'll be hitting the wrong note if you think this is musically inspired: it's actually a reference to a member of staff found on many rail services, given the pub's located right next door to City Thameslink.

Conquering Hero *262 Beulah Hill, Norwood, SE19 3HF*
There are two competing explanations behind this one: they're both great so choose your favourite. One contends it's after the Italian revolutionary Garibaldi, who'd visited Norwood in 1864 a year before the pub opened, and a postcard records he was greeted by the Handel composition 'Enter the Conquering Hero.' The alternative is it was victorious landlord George Hunt, expressing his glee after succeeding against the local gentry who tried to block his licence to open a pub. If it were the latter then it was definitely pride before a fall, as family rivals of Hunt's (they had operated venues next to each other up the road at Beulah Spa) opened a pub next door a year later. The construction of the rival pub weakened the foundations of the Hero, causing it to flood. In 1871 Hunt admitted defeat and the pub was bought by his next-door rival who promptly shut down his establishment, having succeeded in pushing him out. These days the pub is home to a conquering hero of a very different variety, a micropig called Francis Bacon who had to be banned from the bar area after finishing people's pints and then butting into the regulars!

Corner Pin *732 High Road, Tottenham, N17 0AG*
Recently reopened as Beavertown Brewery's first pub after a long period of closure, this name refers to the hardest pin to knock down in a game of skittles.

Coronation Hall *St Mark's Hill, Surbiton KT6 4LQ*
Began its life as a lecture hall which opened the day before George V's coronation, hence the name. It was converted into a pub in 1997, before then it went through various uses from music hall to cinema and bingo hall and even a spell as nudist club according to one online source!

The Coronet *338-346 Holloway Road, N7 6NJ*
This Wetherspoons pub is named after the cinema that once operated on this site as late as the early 1980s. It originally opened as a Savoy cinema in 1940 and with an interlude as the ABC, became the Coronet in 1979. (See The Capitol).

The Crane *14 Armoury Way, SW18 1EZ*
There are two competing theories behind the name of the self-proclaimed oldest pub in Wandsworth. One references its location close to the river and the once common sight of large cranes connected with the heavy industry once based on the river. The second is that it is named after the type of bird, also fond of rivers.

Crocker's Folly.

Crocker's Folly *24 Aberdeen Place, St John's Wood, NW8 8JR*
Now converted into a restaurant, Crocker's Folly had a rather elaborate
pub interior which hints at the legend behind the name. The original
owner was Frank Crocker who was told that the terminus for the new
Great Central Railway was going to be where the pub now stands.
Building an elaborate pub to greet the passengers, he was ruined to find
out the terminus was to be built about half a mile away in Marylebone.
This allegedly led to him committing suicide and his ghost still haunts
the place to this day. However, this is something of a myth as he died
in 1904 of natural causes.

Crooked Billet *Various - Six Pubs in London*
At least two pubs in London have this name. A billet is an old name for
a stick or wooden pole; given it was misshapen, it would often make a
good pub sign and the name has stuck in certain locations, including
Clapton and Wimbledon.

The Cross Keys *Various - Five Pubs in London*
A number of pubs in London are called Cross Keys which is the symbol
of St. Peter, The first Pope and the gatekeeper of heaven. I'm sure the
pub is heaven to many who are reading this!

The Crossing *73 White Hart Lane, Barnes, SW13 0PW*
Located right beside a railway level crossing, the pub nails its colours
to the mast with this name, adopted when it reopened in Autumn 2021.
While the crossing barriers being down might be an inconvenience to
those wishing to pass, at least pedestrians have a pub to pop into!

Crown and Shuttle *226 Shoreditch High Street, E1 6PJ*
Featured in the first edition of Paul Talling's *Derelict London* in a rather sorry state, this pub opened up again in 2013 after 12 years of dereliction; its name is a reference both to the fact lots of the land round here was once owned by the crown and the shuttle being a reference to the silk weaving of Spitalfields.

Crown and Treaty *90 Oxford Road, Uxbridge, UB8 1LU*
A great one for history buffs, this pub references the fact the building was used for negotiations between Charles I and the parliamentarians at the height of the English Civil War in 1645. The wood panelling of the Treaty room (although no treaty was signed here) has had an interesting recent history: in the 1920s it was bought by American billionaire Armand Hammer who transported it to his office on the 78th floor of the brand new Empire State Building. Before taking up the throne, the then Princess Elizabeth visited New York and saw the room, taking an interest in it. When she became Queen, Mr Hammer sent it back to London as a coronation gift and she in turn returned it to the pub.

Crumpled Horn *33-37 Corbets Tey Road, Upminster, RM14 2AJ*
When this pub opened at the turn of the millennium its owners let the local community put forward ideas before the final name was chosen. Other options in the running included The Dury Arms, after the late punk/new wave singer Ian Dury who lived in the area and which would have given his fans a reason to be cheerful. In the end, though, the name of a former local dairy was chosen, from one pint to another.

Curtains Up *28a Comeragh Road, W14 9HR*
This name was adopted in 1993 and references the fact the pub has a small 57-seat theatre in its basement, which in turn pulled down the curtain on its long stint as the Barons Ale House.

Cutty Sark *4-6 Ballast Quay, Greenwich, SE10 9PD*
The pub is, of course, named after the famous ship which sits a short distance away. Cutty Sark was actually the nickname of the witch in Robert Burn's Poem Tam O'Shanter. Apparently the owner of the ship named it after a witch as they are unable to cross water in the poem.

Dame Alice Owen *292 St John Street, EC1V 4PA*
Alice Owen was born in the mid-16th Century and lived until 1613. She married three times in her life, surviving all of them. Upon the death of her third husband, she went into philanthropy and built a number of almshouses for fellow widows. She also opened a school which still bears her name to this day. Her father was an Islington innkeeper and her first husband a master brewer, which makes it rather fitting for her to be commemorated in this way.

Danson Stables *Danson Park, Crook Log, DA6 8HL*
As the name suggests, this pub is based in what was the old stable block for Danson House and dates back to the start of the 19th century. The House itself was completed in 1768 for Sir John Boyd, owner of the Danson Estate.

Daylight Inn *Station Square, Petts Wood, BR5 1LZ*
This commemorates that the first person in the UK to actively call for the introduction of 'Daylight Saving Time', William Willett, lived in Petts Wood. Sadly, he died in 1915, a year before its first introduction in Britain, albeit in May as opposed to March. There is a memorial sundial to Willett in Petts Wood and which always gives its readings in British Summer Time, a bit like a rogue household appliance where the timer simply won't change despite your best efforts.

De Hems *11 Macclesfield Street, West End, W1D 5BW*
De Hems claims to be the only Dutch themed pub in London. It was certainly popular with Dutch sailors and during World War II became a key location for the Dutch resistance. The name comes from one of the landlords during the 19th Century, who was a retired sea captain and ran the place as an oyster house. The upstairs walls were once adorned with 300,000 oyster shells.

Dean Swift *10 Gainsford Street, Southwark SE1 2NE*
A craft beer pub near Tower Bridge, it's named after Johnathan Swift, the Irish essayist and poet who is best remembered for writing *Gulliver's Travels*. Dean is in fact his nickname, after he became the Dean of St. Patrick's Cathedral in Dublin.

Deers Rest *Noah Hill Road, RM3 7LL*
For much of its life this pub was known as The Bear and for many years they even had a caged bear on the premises. Those days are thankfully long gone and when the pub was refurbished in 2018 the owners decided they wanted to break with that past and pick a new name to reflect the local area. The pub is surrounded by one of the UK's largest roaming deer herds, which can often be seen in the grounds grazing or, as the pub name suggests, resting!

Defector's Weld *170 Uxbridge Road, Shepherd's Bush, W12 8AA*
A slightly cryptic name but rather fitting when you discover the meaning. It's said to be a reference to the fact one of the notorious 'Cambridge five' (spies who were passing information to the Soviet Union from the 1930s through to the 1950s) worked at the BBC just up the road at Wood Lane, hence the Defector. The Weld is for the act of joining or coming together. Perhaps appropriately we can find no evidence of them drinking here...

Dirty Dicks.

Depot *7 Peglar Square, SE3 9FW*
This pub in a new development takes its name from the citing of an
RAF station here, which was used to store barrage balloons.

Dial Arch *Riverside, The Warren Royal Arsenal, No 1 Street, Woolwich,
SE18 6GH*
This pub is based by the front entrance to the Dial Square which
is entered through passing the Dial Arch, hence the name. In 1886
workers in the Dial Square formed a football club which they soon
called Royal Arsenal before becoming Woolwich Arsenal. In 1913
they moved to Highbury, dropped Woolwich from their name (as that
would just be confusing..) and the rest is history!

Dirty Dicks *202 Bishopsgate, Spitalfields, EC2M 4NR*
Certainly one of the cheekier named pubs in London, this pub is named
in honour of the former owner and merchant, Nathaniel Bentley. His
story is rather tragic as he picked up the name Dirty Dick after refusing
to wash after the death of his fiancée on their wedding day. This led to
living in squalor for the rest of his days and his house and warehouse
became famous for its disgusting state. The pub's cellar used to pay
homage to him but, thankfully, it has now been tidied up, but a display
cabinet still has some items on display.

Distillers *64 Fulham Palace Road, W6 9PH*
Named after the Hammersmith Distillery which was based nearby and
opened in 1857. The Distillery building was demolished in the 1990s
and replaced by the Fulham Reach apartment complex.

Dodo Micropub *52 Boston Road, W7 3TR*
There are multiple layers to the name here, one being a nod to the family name of its creator, Lucy Do, another being taking things back to an earlier era of small, community-based pubs before the growth of larger establishments tailored to either sports or food. Finally, Lucy's previous career was in marketing and she had a strong vision of an illustrated dodo being part of the pub's branding. And, quite frankly, why not?!

The Dog House *293 Kennington Road, SE11 6BY*
A humorous name for this dog-friendly pub. Obviously it makes quite a good story if you're 'in the dog house' when having a pint!

Doggett's Coat & Badge *1 Blackfriars Bridge, SE1 9UD*
Thomas Doggett was an Irish actor responsible for establishing the world's oldest boat race from London Bridge to Chelsea between two pubs called The Swan, with the first edition in 1715. Legend has it that he fell into the Thames and was rescued by several watermen (who used to row people across the river). He then established a race as a bet to see who was fastest, with the winner receiving a traditional waterman's red coat with a badge which would honour King George I, who had ascended the throne the year before Doggett fell into the river. The race continues each year to this day.

The Doric Arch.

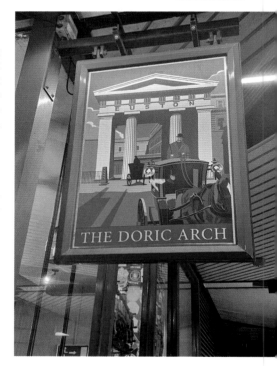

Door Hinge *11 Welling High Street, DA16 1TR*
London's first micropub, this Welling establishment is named after the first landlord's mother. Of course, she wasn't called Door Hinge: her maiden name was Doreen Indge and her fellow pupils shortened Doreen to Door which in turn led to Door Hinge!

The Doric Arch *1 Eversholt Street, Euston Square, NW1 1DN*
Demolished as part of the redevelopment of Euston station in the 1960s, despite the protestations of John Betjemin, there was a Romanesque arch that used to sit outside the station in a Doric style. The Fuller's pub that is now on the site is named in honour of the arch. It was previously a canteen for London Underground staff and by the 1990s it was called The Rails, putting on live bands which included Green Day's first UK gig in 1991.

Douglas Fir *144 Anerley Road, SE20 8DL*
This Gipsy Hill Brewing Company pub first opened in 2016 and its wooden bar is largely made up of this type of pine tree. The owners acquired it from the pub's neighbours, a high-end flooring company who had some surplus fir left over, which not only furnished the bar but named the pub!

Duchess Belle *101 Battersea Park Road, Nine Elms, SW8 4DS*
Formerly called the Duchess of York, when the pub was taken over by Belle Pubs they kept the first part of the name and added theirs. They've done the same for their other pubs.

Duchy Arms *63 Sancroft Street, Kennington, SE11 5UG*
A nod to the fact that a significant proportion of the Kennington area used to be owned by the Duchy of Cornwall's (Prince Charles) estate. The lion's share was sold off to a housing association in 1990 but the Duchy still retains just under 40 properties in the area.

Dutch House *2 Mercury Way, SE12 9AY*
This one is a bit of 'does what it says on the tin' name, as the pub building here was designed specifically to look like a Dutch House. The brewery even sent local architects from Woolwich over to the Netherlands so they could see first-hand the style they wanted them to replicate for this imposing 1930s road house pub. The Dutch narrowly survived being turned into a fast food outlet in 2016 and reopened in spring 2021, repurposed as an all-day operation, doing a café style breakfast and lunch operation during the day before returning more to its roots with more of a pub vibe as the day goes on.

Earl Beatty *365 West Barnes Lane, KT3 6FJ*
This pub opened in 1937, a year after the death of David Beatty, a Naval commander during the First World War and then subsequently First Sea Lord (head of the Royal Navy) between 1919-27.

Earl of Chatham *15 Thomas Street, Woolwich, SE18 6HU*
A Prime Ministerial pub here, in the sense that is named after William Pitt the Elder, who would take up the title of Earl of Chatham in 1766 when he became PM. He would be in post for two years and, of course, his son would become Prime Minister himself in the 1780s.

Earl Ferrers *22 Elora Road, SW16 9JF*
Thought to be named after Laurence Shirley, 4th Earl of Ferrers, the last nobleman in England to be executed, with a painting of Shirley contained within the pub. There is some speculation this connection has been made retrospectively, given the pub opened in the late Victorian era, well over a century after Shirley's execution, and the fact it is located on the corner of Ferrers Road, also from that era and not thought to be named after Laurence.

Earl Spencer *260-262 Merton Road, SW18 5JL*
John Spencer, the first Earl Spencer, acquired vast tracts of land in what is now Wandsworth in the 18th century. It remained in his family until the mid-19th century when significant parcels were sold off as housing rapidly expanded across Battersea, Wandsworth and Southfields. As well as this pub, there is another in Putney simply called The Spencer, another area where the family owned significant areas of land.

East India Arms *67 Fenchurch Street, EC3M 4BR*
This City pub is named after the East India Company, whose headquarters used to be next door to the pub. The pub only opened for the last 20 years of the company's 270 years existence but, given it was next door, it's not hard to imagine it became a favourite watering hole before the company's end.

Edgar Wallace *40-41 Essex Street, Temple, WC2R 3JE*
In a book full of pubs named after characters or stories, here we have a venue named after a character who wrote stories, lots of stories... Edgar Wallace was a truly prolific novelist of the early 20th century, churning out an incredible 170 novels during his life and 957 short stories. It was rumoured that part of the reason he produced so many books was to cover his gambling debts. His connection with the area is that one of his very early jobs during his teens was as a newspaper seller on nearby Fleet Street and he had a stint working as a reporter for the *Daily Mail* before becoming such a prolific writer.

Editor's Tap *5-11 Fetter Lane, EC4A 1BR*
A nod to the area's former life as the heart of the UK newspaper industry.

The Edmund Halley *25-27 Burnt Ash Road, SE12 8SS*
Edmund Halley was a famous astronomer whose work in observing transits of Mercury and Venus helped prove Newton's Laws of Motion which was instrumental in advancing our knowledge of the solar system. He also discovered that the same comet was appearing in the sky every 75 years and it was subsequently named after him,. although he sadly passed away before he could see it appear in the sky. The pub is near the Greenwich Observatory, which might explain why Wetherspoons named the pub in his honour.

Eel Pie *9-11 Church Street, TW1 3NJ*
This is not named after any delicacies on the pub's menu but instead references an island on the Thames opposite the pub. In its heyday in the 1960s, the hotel on the island saw huge names such as Pink Floyd, the Rolling Stones and The Who all play there on their way to fame. Sadly, the venue closed in 1967, briefly reopening in 1969 under a different name but still managing to add to its impressive roll-call with acts such as Black Sabbath and Genesis. The hotel burnt down in 1971 but the creative spirit on the island lives on with various artists who base themselves here. They open up their studios a few times a year if you're feeling curious.

Effra Social *89 Effra Road, SW2 1DF*
Just down the road from Brixton, this pub used to be a social club, hence the name. The Effra is one of London's subterranean rivers which runs underground in the area.

Eight Bells *89 Fulham High Street, SW6 3JS*
Marks the fact that in 1729 Fulham Church acquired two extra bells and became an eight belled church. Was briefly used as changing rooms for Fulham FC between 1886 and 1888 when they played at nearby Ranelagh House, one of many early homes for the club before they settled at Craven Cottage in 1896.

Eleanor Arms *460 Old Ford Road, E3 5JP*
Named after Eleanor of Castile, Edward I's wife, as it is thought her funeral procession passed through this area on its journey from Waltham Cross to Charing Cross.

Elephant & Castle *Various - Three Pubs in London*
While there is the famous pub near the tube station of the same name, there are also several others across the city. The short story of this is that the Cutler's Company (one of the City of London's livery companies) has both an elephant and a castle in its crest. There is a popular myth that it comes from the Spanish phrase 'Infanta de Castile' which is the title given to the oldest daughter of the Castile monarch.

Elephant's Head.

Elephants Head *224 Camden High Street, NW1 8QR*
An elephant's head features on the coat of arms of the Marquess of
Camden and subsequently featured on the coat of arms for the London
Borough of Camden.

Eltham GPO *4 Passey Place, Eltham, SE9 5DQ*
This name is a nod to the building's original use as a Royal Mail sorting
office, GPO standing for General Post Office for those not old enough
to remember when the postal service had that name. This is the second
incarnation of a pub on this site, the first version was simply called the
Old Post Office.

Enderby House.

The Faltering Fullback.

Enderby House *23 Telegraph Avenue, SE10 0TH*
This pub is located within Enderby House, first owned by Samuel Enderby & Sons, Britain's largest whaling company during the first half of the 19th century. They also lent their name to this entire wharf area of Greenwich.

Euston Tap *190 Euston Road, NW1 2EF*
Named after the mainline station that sits behind, the Euston Tap sits in two old gatehouses that are the last remnants of the original station. Euston comes from the seat of the Duke of Grafton, Euston Hall, in Suffolk, who owned the land adjacent to the station.

Eva Hart *1128 High Road, Chadwell Heath, RM6 4AH*
Named after one of Chadwell Heath's most famous residents, one of the oldest Titanic survivors, who died in 1996 having reached the ripe old age of 91.

The Faltering Fullback *19 Perth Road, Finsbury Park, N4 3HB*
Fullback is one of the playing positions in rugby. One can imagine them faltering when catching the ball. The pub pays homage to its name by having sport on big screens and has lots of rugby memorabilia on display. It also has a multi-levelled beer garden that looks like the Ewok Village from Star Wars and a Thai kitchen.

Famous 3 Kings *173 North End Road, W14 9NL*
Originally known as the Three Kings, it had a spell as during the 70s as a live music venue called the Nashville, hosting the Sex Pistols amongst others. It was rechristened the Famous 3 Kings in 2000. This

The Fellow.

time around the famous three were Henry VIII, Charles I and Elvis, a combination which feels like an easy multiple-choice question on the itbox machines that were once a staple in pubs across the country.

Farr's School of Dancing *17-19 Dalston Lane, E8 3DF*
This pub used to be a commercial premises and was originally a dance studio back in the 1930s.

Fat Walrus *44 Lewisham Way, SE14 6NP*
This is a reference to the Horniman Museum's most famous exhibit, a large stuffed walrus. The large mammal has been based at the museum since the tail end of the 19th century, having originally been brought to Britain as part of an exhibition in South Kensington in 1886. This pub isn't actually that close to the Horniman; the owners had originally had their eye on somewhere much closer to the museum in Forest Hill but when that site fell through, still decided to persevere with the name.

Fellow *24 York Way, N1 9AA*
This pub galloped through various names over the years before being rechristened in 2008 after a successful racehorse from the 1990s whose honours included the 1994 Cheltenham Gold Cup. This name was chosen as a picture of this horse was hanging up in the pub when the then owners acquired it and they decided it was clearly a winning bet.

Fellowship & Star *Randlesdown Road, SE6 3BT*
Built to serve the Bellingham Estate nearby, the Fellowship was one of the first pubs built by the London County Council after the Temperance Movement had successfully kept the pubs off their earlier estates. This large pub was popular with bands and included a theatre on the premises. After securing a Lottery grant in 2016, these spaces were refurbished and the Electric Star Group would run the pub, hence the addition to the pub's name.

The Festival inn.

Fence *67-69 Cowcross Street, EC1M 6BP*
The name here reflects the surprising fact about this pub in the heart of built-up Farringdon, it's got a large garden at the back!

Ferry House *26 Ferry Street, Isle of Dogs, E14 3DT*
Prior to the opening of the Greenwich foot tunnel in 1902, the only way to pass from the Isle of Dogs to south of the river was by ferry, which operated right by the pub, hence the name!

Festival Inn *72 Kerbey Street, Poplar, E14 6AW*
Opened in 1951, the year of the Festival of Britain, alongside Chrisp Street Market, both designed by renowned architect Frederick Gibberd, who was closely linked to the immediate post war reconstruction of Britain through his work on Harlow New Town, including Britain's first tower block, the Lawn.

Fig Tree *49 Windsor Street, Uxbridge, UB8 1AB*
This building was once used as a police station, something referenced in the original pub name here of Old Bill. However, the owners decided they wanted to break free from the past and renamed it after the tree in the pub's garden. While you might suspect this to be a common pub name, this is actually the only one in Britain!

Finch's *12A Finsbury Square, EC2A 1AN*
Honours the first owner of the pub on this site, Henry Hobson Finch. Finch's brewery ended up owning countless pubs across London, notable names including Dirty Dicks, and existed up until 1991 when the business was sold to Youngs, who have kept his name alive here.

Fire Stables *27-29 Church Street, Wimbledon, SW19 5DQ*
Marks the fact that the old fire station just up the road here used to
have stables at the back for the horses that used to pull the fire engines
before the arrival of the car.

Fishmongers Arms *Winchmore Hill Road, N14 6AD*
A pub with this name first arrived on this site in 1688. The name is
thought to derive from the fact there were ice wells situated at the back
of the building; this ice was then transported into London for use in
the fish markets and associated market stalls.

The Flask *14 Flask Walk, Hampstead, NW3 1HE*
and 77 Highgate West Hill, N6 6BU
In the introduction we stated we wouldn't include pubs named after
the street they're on, which on first glance, seems to rule out this
Hampstead pub; however, here it's the other way round and the street
was named after the pub. Its origins date from the start of the 18th
century, when water from the wells of Hampstead began being sold
for medicinal purposes. The water was collected up, taken to this pub,
then called The Thatched Inn, where it was then bottled up in flasks
and sent across London. This practice is also the reason behind the
pub of the same name at the other side of the Heath, in Highgate.

The Florin *563 Holloway Road, Archway, N19 4DQ*
A Florin was an old English coin worth two shillings and was replacedf
in 1971. It picked up its name as it was stamped with the image of a lily
and its floral design gave the coin its name.

Flowers of the Forest *14 Westminster Bridge Road, SE1 7QX*
Said to take its name from a play (*Flower of the Forest*) that was
showing at the adjacent Surrey Theatre in 1845 and led the landlord
of the "Surrey Coal Hole' to adopt the play's title for his pub. It wasn't
actually this pub he renamed, the original Flowers of the Forest pub
was demolished in the early 1990s but the name was transferred over
to this pub, then called the Oxford Arms, a short distance from the
previous location.

Flowerhouse *56 Blandford Street, Marylebone, W1U 7JA*
A tribute to the flower stall that was based outside the pub for
many years and served the good folk of Marylebone for their floral
requirements.

Flying Chariot *Heathrow Terminal 2, TW6 1EW*
Very apt name for an airport pub, this comes from the 17th century
verger Dr John Wilkins, local to these parts, who wrote about possible
voyages to the Moon in a Flying Chariot. It certainly has a nice ring
to it so Dr Wilkins might have been onto something. This pub is pre-
security, so you don't need to have a flight to drop in.

Flying Machine *79 Kings Road, Biggin Hill, TN16 3XY*
Closest pub to Biggin Hill Airport on the outskirts of London, a pivotal site in the Second World War. It's still a registered airport today, but is primarily used for small private flights and flying lessons, as well as summer air shows.

Footman *5 Charles Street, Mayfair, W1J 5DF*
When it was known by its earlier and longer name of I Was The Only Running Footman, it had the honour of having the longest pub name in the capital. It has since been shortened twice, first to The Only Running Footman and then to its present name in the last decade. Regardless of the tinkering, it already lost its crown in the 1980s as a result of the growth in 'quirky' pub names. It comes from the old profession of Footmen, people who would clear the path ahead for the horse drawn carriages of the 18th century. The pub was bought by a footman who renamed it from its original name The Running Horse. In 1986 the author Martha Grimes wrote a crime novel named after the pub (during the I Was era) with the pub playing an integral role in the story.

Founders Arms *52 Hopton Street, SE1 9JH*
Named after Founders Wharf, which was once located beside here on the South Bank. The current pub building dates from 1980 after the original was demolished in the 1960s after getting badly damaged during the blitz.

Founders Arms.

Fountains Abbey
109 Praed Street, W2 1RL
This name harks back to the character of Paddington in days long past, the 'Fountains' refers to the fact there were once springs and mineral wells in the area, the 'Abbey' for the fact that this part of Paddington used to lay within the manor of Westminster Abbey. The pub is opposite St Mary's Hospital and it is claimed that it was mould spores from here than drifted over the road and into Alexander Fleming's lab and led to his discovery of penicillin. If that is the case, then this pub really has done all of us a favour over the last 90 plus years!

Fountain and Ink *52-54 Stamford Street, SE1 9LX*
Takes its name from the fact the pub is housed in the first registered headquarters of Stephen's ink company, set up by Dr Henry Stephens who in 1832 invented his indelible 'blue-black writing fluid' which

French House.

Friend at Hand.

became famous as Stephens Ink. The ink was used in the pens that signed the Versailles Treaty in 1919 and, although the company name disappeared in 1967 following a takeover, the ink is still mandated to be used on official registrars' documents in the UK such as birth, marriage and death certificates.

Four Quarters *187 Rye Lane, SE15 4TW*
and 20 Ash Avenue, SE17 1GQ
This pub is packed full of retro arcade games, hence the four quarters. After opening their first in Peckham in 2014, they advanced to the next level when they opened their second venue in Elephant and Castle.

Four Thieves *51 Lavender Gardens, Battersea, SW11 5DJ*
Named after a tale from medieval Britain during the height of the black death where four corpse robbers were saved from that same grisly fate by lavender (whose former presence in this area is referenced in the road name) which made them immune to the plague.

Foxglove *209-211 Liverpool Road, N1 1LX*
When the new owners took over here, they were keen to name it after a song title from one of their favourite bands, Murder by Death. After deciding most of their song titles weren't suitable, they decided on Foxglove. In telling us the story, Kimberley, one of the owners, said another reason the name stuck was that, as her business partner said, it would be a good fit because, like her, "it's pretty but it will kill you!".

Foxley Hatch *8-9 Russell Hill Parade, Russell Hill Road, CR8 2LE*
Named after the Foxleyhatch Turnpike (a toll road in today's terms) which used to stand in this area.

The French House *49 Dean Street, Soho, W1D 5BG*
Formerly called the York Minster, this tiny Soho pub became an unofficial base for the French Resistance during the Second World War. Charles de Gaulle himself and his generals are known to have drunk here whilst working on their plans to retake France and thus the name was changed. The pub is also popular amongst London's creative community.

Friend at Hand *2-4 Herbrand Street, WC1N 1HX*
The pub sign here depicts a dog and you'd assume that would be the 'Friend' being referred to. Looking back at its history, when the pub first opened in 1797 it was called the Handsome Carriage and based close to the 'Horse Hospital', stabling facilities for carriagemen's sick horses. It's not an impossible stretch to see horses being the original inspiration for the name, as they could also be considered a 'Friend at Hand'.

Furze Wren *6 Broadway Square, Market Place, Bexleyheath, DA6 7DY*
Takes its name from a bird which was discovered for the first time on Bexleyheath by a Doctor John Latham in 1773. The bird's name comes from the fact it likes to nest in furze bushes. It is also referred to as the Dartford Warbler, which would also make for a fine pub name.

Gallery *1 Lupus Street, SW1V 3AS*
The picture they're trying to paint here is that this pub is close to Tate Britain, although it isn't the nearest to the gallery: we suppose the owners at the time used their own version of artistic licence.

Galvanisers Union *2 Devas Street, Bromley-by-Bow, E3 3LL*
This recalls the former presence of the Lancaster and Co galvanising works factory on Hancock Road, a short distance from the pub.

Ganymede *139 Ebury Street, SW1W 9QU*
Named after Jupiter's largest moon, due to the fact Ganymede in Greek mythology was a Trojan aristocrat kidnapped by Zeus and taken to Mount Olympus to become 'the Wine Pourer/cupbearer to the Gods for all eternity.' The reason the owners felt this was an apt name for the pub was due to the fact it was the Ebury Wine Bar in its last incarnation, hence the Ganymede connection.

Garden Shed *145 Haydons Road, Wimbledon, SW19 1AN*
This name was picked via an online vote when this pub reopened in late 2013, having previously been known as the Horse and Groom. The pub has a large garden which we think was the inspiration behind this.

Garrick Arms *8-10 Charing Cross Road, WC2H 0HG*
Takes its name from the nearby theatre which opened in 1889.

Garrison *99-101 Bermondsey Street, SE1 3XB*
Renamed in 2003 when new owners took over, Garrison was chosen as
a way of demonstrating you'll find the pub as a safe, welcoming place
protected from the outside world.

The Gate Clock *Cutty Sark Station, 210 Creek Road, Greenwich, SE10
9RB*
The actual clock from which the pub gets its name is found on the gate
of the nearby Greenwich Observatory, which is the official source of
Greenwich Meantime (GMT).

General Roy Poplar Way, Feltham, TW13 7AB
The General this pub refers to is William Roy, a military engineer and
surveyor in the late 18th century who developed the first accurate
land measurements in 1784, realising the importance from a military
perspective of knowing exactly where things are, which eventually
led to the establishment of Ordnance Survey. He began his task by
establishing the precise positions of Greenwich and Paris observatories
and decided to use the method of triangulation, which works by setting
out a baseline, measuring the angle of a third point from both ends,
and a network of interlocking triangles would eventually pinpoint the
location of everywhere on the grid. And the connection to Feltham?
For Roy's initial baseline he needed two endpoints five miles apart
with a large, flat expanse in between: he chose Hounslow Heath, with
the five-mile baseline running through Feltham itself.

The George *Various, Seven Pubs in London*
Of the seven pubs in London with this name, the most famous is
unquestionably the one on Borough High Street. It began life as The
George and Dragon, so is named after St George as opposed to one of
the six British Kings with that name. There is so much you could write
about this pub that you could easily fill an entire book, and the excellent
writer Pete Brown has done just that with his book Shakespeare's
Local. It is also the only pub in London owned by the National Trust
and the last remaining galleried coaching inn, a reminder of the days
where people would have stopped here for the night before setting off
on their journey.

George & Dragon *Various - Six Pubs in London*
England's patron saint is at the centre of a legend where he fought and
slayed a dragon in order to save a maiden. Many pubs honour this but
George died in 303 AD and the legend didn't surface until the 1300s, so
the factual accuracy of this can be somewhat questioned!

George and Monkey *68 Amwell Street, EC1R 1UU*
In recent years this place has gone through multiple names: after
shedding its long-time name of The Fountain to become Filthy
McNasty's, it then turned into a Simmons cocktail bar before having

The George.

a brief stint as the Amwell Arms. The current name comes from the previous owners of the pub, who named it after their son George who was a fan of monkeys. They've since moved on but the name has stuck, which makes a change for this place.

George & Vulture *3 Castle Court, EC3V 9DL and 63 Pitfield Street, N1 6BU*
There are two George & Vultures in London. The oldest is located in the Square Mile and dates from the 1470s when it was originally called The George. The vulture came after it was rebuilt following the Great Fire in 1666. The landlord kept an actual vulture above the door until the name stuck. He later saw this wasn't great for business and let the beast fly away. Lately this pub has many connections to Dickens who regularly drank here and was where Mr Pickwick resided in The Pickwick Papers. The second one is found north of Old Street and is London's tallest pub as well as being a favourite of George Orwell, who wrote an essay on it in 1946.

George Staples *273 Blackfen Road, DA15 8PR*
Was originally the Woodman until being renamed in 2008 in honour of the first landlord of the pub in 1845, who himself had named it after his own trade.

Gertie Brownes *95 High Road, East Finchley, N2 8AG*
This inn is named after the pub in Athlone, Ireland where the couple who own the pub first met. They had to ask permission to use the name but otherwise there's no connection.

Gilpin's Bell *50-54 Fore Street, N18 2SS*
Takes its name from the protagonist of a comic ballad written in 1782 by the poet William Cowper, 'The Diverting History of John Gilpin'. In Cowper's tale, Gilpin was a wealthy draper whose attempts to reach The Bell Inn on Fore Street in Edmonton, to celebrate his 20th wedding anniversary with his wife, were calamitous. His first attempt saw him unable to control his horse which kept on going past The Bell to end up in Ware, 10 miles away. Trying to get back to The Bell, the horse wouldn't stop again and took him right back to where he started, in Cheapside in the City. Various pubs in the area claimed to be The Bell from the poem, sadly none of them now exist. This pub is relatively modern, dating from the late 1990s. As well as being commemorated in the pub name there is a stone statue of a Bell, carved with images from Gilpin's disastrous ride, also on Fore Street.

Gipsy Moth *60 Greenwich Church Street, SE10 9BL*
Originally called The Wheatsheaf, was renamed in 1975 in honour of
Sir Francis Chichester's Gipsy Moth IV ship which he sailed single-
handedly across the globe in 1966-67. A year later it was put on
permanent display in Greenwich beside the Cutty Sark where it still
stands to this day.

The Globe.

The Good Samaritan.

Glassblower *42 Glasshouse Street, W1B 5JY*
A reflection of the fact this site was once used by glassblowers It was
also a workhouse at one point, but that would make for a rather bleak
pub name.

The Globe *Various - Seven Pubs in London*
This name comes from the globe being used as the emblem of Portugal
and advertised the fact that fine Portuguese wines were on sale. There
have been many more pubs in London called The Globe, peaking at
over 30 in the 19th Century. The Globe in Moorgate dates from the time
of Charles I and the one in Borough Market was famously used in the
movie Bridget Jones's Diary, with her flat being above the pub.

Goat Tavern *3 Stafford Street, W1S 4RP*
Thought to be linked to the Worshipful Company of Cordwainers, a
city livery company covering those working in the leather trade and
whose coat of arms contained a white goat. There is another goat
nearby in Kensington; whether or not that is connected to the livery
company is unknown.

The Goldengrove *146-148 The Grove, E15 1NS*
The famous poet, Gerard Manley Hopkins, was born in Stratford where this pub is found. Goldengrove is included in a line of his poem, 'Spring & Fall'. (See also The Skylark).

Golden Ark *186 Addington Road, Croydon, CR2 8LB*
The micropub's name was picked to symbolise the veritable bounty of beers from across Europe and beyond they have up on offer, like stumbling on a Golden Ark!

Golden Crane *117 Avon Road, Upminster, RM14 1RQ*
Here we have a name born out of misunderstanding: in short, a belief that Cranham near Romford is named after the hunting of cranes. In reality, it derives from a word that means land frequented by crows.

Golden Fleece *166 Capel Road, Manor Park, E12 5DB*
and 9 Queen Street, EC4N 1SP
We go back to ancient Greece for this pub name. The golden fleece was made from ram's wool and would have been used to collect gold particles by stretching them across the river. It was then stolen by Jason and the Argonauts as he believed it would place him on the throne, as the fleece was a symbol of authority.

Gold Coast *224 Portland Road, London SE25 4QB*
Acquired its present name in 2004 after being taken over by a team with Ghanian roots, this pub now serves up a West African twist on the Gastropub concept.

The Good Mixer *30 Inverness Street, Camden Town, NW1 7HJ*
Famous for being the drinking hole for many famous musicians in the 1990s Britpop era, The Good Mixer takes its name from a cement mixer that was accidentally trapped in the foundations during the pub's construction in the 1950s. It is also where the Battle of Britpop between Oasis and Blur began and in the 00s Amy Winehouse was a regular customer.

Good Samaritan *87 Turner Street, E1 2AE*
This early 19th century pub is located beside the Royal London Hospital, which is thought to be the inspiration for its name. The Hospital has had a representation of the City of London as a good Samaritan on its official deal since 1757.

The Good Yarn *132 High Street, Uxbridge, UB8 1JX*
The building which the pub occupies used to be Pearson's menswear shop. The company was founded as a tailor's in the 1830s using good yarn to make their suits.

Goodman's Fields *87-91 Mansell Street, E1 8AN*
The area this pub now stands in was once known as Goodman's Field, named after Roland Goodman who farmed the land for the nuns of the nearby St Clare's Abbey.

Gorringe Park *29 London Road, Tooting, SW17 9JR*
Named after a 19th century villa called Gorringe Park which once stood in this area. Time was called on the villa when it was demolished at the start of the 1930s.

The Graduate *107-109 Blackheath Road, Greenwich, SE10 8PD*
In a previous life this was called the Coach and Horses; it was renamed at the turn of the millennium by its new owner who wanted to reflect its proximity to Goldsmiths University. He narrowed it down to two names, either The Graduate or The Scholar. Testing the names with friends back home in Ireland, they thought it was The Graduate which made the grade and the rest is history.

The Grapes *76 Narrow Street, Limehouse, E14 8BP*
Any pub referring to grapes can usually trace its name back to Roman times when wine was the tipple of choice. Landlords used to display a bunch of grapes on the sign outside to indicate that fresh wine was available and this tradition continued. Arguably, the most famous pub with this name is the one in Limehouse, which dated back nearly 500 years. It has been frequented by Pepys and Dickens and has inspired Oscar Wilde and Arthur Conan Doyle. It is currently owned by Sir Ian McKellen, who saved the pub from closing down. A subtle hint that he's the owner is the presence of Gandalf's staff from *The Lord of the Rings* movies, which sits behind the bar.

Great Exhibition *193 Crystal Palace Road, SE22 9EP*
Named after the spectacular event in 1851 housed within the Crystal Palace in Hyde Park. After the exhibition ended that October, the palace was dismantled and rebuilt in the Sydenham Hill area of south London; however, such was the draw of this unique building, the whole area soon ended up being named after it.

The Great Harry *7-9 Wellington Street, SE18 6PQ*
Great Harry was the nickname of the flagship of King Henry VIII. Officially called 'Henry Grace à Dieu' which translates as 'Henry thanks be to God', it was the largest ship in the world at the time, even larger than its more famous contemporary, the Mary Rose. It was built at Woolwich Dockyard which was founded for its construction. The docks would go on to build other ships (see The Sovereign of the Seas).

The Great Northern Railway Tavern *67 High Street, Hornsey, N8 7QB*
Located on Hornsey High Street near to Hornsey train station, it is named after the former company which built King's Cross station and

The Green Goose.

ran services up to Doncaster, where it allied with the North Eastern Railway to allow services up to Scotland. They were both merged into the London North Eastern Railway in 1923.

The Great Southern *79 Gipsy Hill, SE19 1QH*
This pub's website proudly proclaims it was named after a steam train. It doesn't state where it ran, though: the closest fit we've been able to find was a set of steam trains called Great Southern Railways which were built not too far from here in Woolwich in the 1920s, then exported to Ireland.

Great Spoon of Ilford *114-116 Cranbrook Road, Ilford, IG1 4LZ*
While this is a Wetherspoons, the 'Great Spoon' in question is not some idle boasting from Tim Martin but reference to a historical event. In 1600 the famous actor Will Kemp danced his way from Norwich to London; the 110-mile journey took him nine days and he stopped in Ilford for refreshment, and the measure of ale was called a spoon (comparable to two pints). Given he'd been dancing all the way, you can only assume he'd had a fair few spoons on the way too…

Green Goose *112 Anglo Road, E3 5HD*
This pub is named after the Bow Fair, which used to take place every Whit Sunday on the site of the Old Poplar Town Hall, and had the nickname Green Goose Fair, in part due to the fact roasted geese were sold there. Green Goose was also the nickname for loose women and the fair built up a reputation for being a lively event. Continuing that theme, we defer to the 'Look Up London' website, who suggest it has London's best memorial plaque: it reads 'Sight of Annual Whitsun Fair stopped in 1823 due to rowdyism and vice'.

The Gregorian *96 Jamaica Road, Bermondsey, SE16 4SQ*
This one posed quite a puzzle to decipher the origins of the name. It was finally revealed in an Evening Standard interview with the then landlord in 1937. He explained it came from The Gregorians, a secret society like the Masons which once met in the pub, but that he now used the room they met in for the pub's billiard table!

The Grenadier *18 Wilton Row, SW1X 7NR*
While this pub is famous for its Bloody Marys, the Grenadier takes its name from the fact the building is the former officer's mess for the Grenadier Guards. Indeed, many of them still drink in the pub to this day.

Greystoke *7 Queens Parade, W5 3HU*
Named after a Victorian mansion house which once stood in this area and was built for John Carver, an industrialist from Lancashire. After his death the building had various owners, including Arthur G Sands, CEO of Walls Ice Cream and Sausages, and the Post and Telegraph Union HQ before being demolished in the late 1970s.

The Griffin *Brook Street South, Brentford, TW8 0NP*
Named after the Fullers brewery, located a few miles east along the river. The Griffin is traditionally a symbol of the City of London, it ended up being the name of a brewery in Chiswick because the first owners acquired the business from a brewery based in the City. There are various other pubs dotted around London called The Griffin, mainly due to the City connection. We've included this one over the others as it gave its name to Griffin Park, Brentford FC's home ground for 114 years. This surely must make it the only pub that has named a football ground.

The Guinea.

Guinea *30 Bruton Place, W1J 6NL*
Two competing explanations here: one that seems fitting with the current character of the area, another which might shock you. First, the high-end version: a Guinea was one pound and one shilling in Britain's pre-decimalisation currency, favoured by well-heeled establishments and those who frequented them, an ideal match for posh Mayfair. The other is somewhat less glamorous. The pub was originally called the pound because of the presence of a cattle and animal pound where Berkeley Square now stands; the name changed as stable boys jokingly rechristened it as the Guinea in 1663, given that at that point a guinea was worth a pound.

Gunmakers *13 Eyre Street Hill, EC1R 5ET*
In 1881, Hiram Maxim, a prolific inventor, set up a factory at nearby Hatton Garden. While Mr Maxim patented countless innovations in his time, his most famous creation was the automatic machine gun and it is from that the pub gained its name. His other inventions included curling irons which could have also led to an intriguing pub name too…

Hackney Carriage *165 Station Road, DA15 7AA*
We'll let you into the knowledge here: this one derives from the fact that the husband of this micropub owner is a cabbie.

Hagen and Hyde *157 Balham High Road, SW12 9AU*
There are two distinctive elements behind this name. The first, Hagen, commemorates a former partner of Anthony Thomas, the owner of the Antic pub group, called Rita Hagen who was from Balham but who sadly died at a young age. Hyde is a reference to Hyde Farm, which once covered a significant area of Balham and was initially owned by Emmanuel College, Cambridge. After the arrival of the railway in the 1850s, it was soon one of the last remaining farmlands in the area until it was developed for housing from 1896 onwards.

Half Moon *213-223 Mile End Road, E1 4AA*
With any Wetherspoons, the immediate assumption when you spot moon in the name is that it's a derivation of their usual formula around 'Moon under Water', George Orwell's favourite yet totally fictitious pub. Here the story is slightly different, as its roots come from the Half Moon Theatre Company which occupied this former Methodist chapel for 13 years from 1977 to 1990.

Halfway to Heaven *7 Duncannon Street, WC2N 4JF*
An LGBT+ venue since the early 1990s, this bar is not too far from the famous Heaven nightclub, making this place a good venue for pre-drinks before a night on the tiles.

Hamilton Hall *Liverpool Street, EC2M 7PY*
Liverpool Street Station's branch of Wetherspoons is named after the chairman of the Great Eastern Railway Company that built the station, Lord Claud Hamilton, who ran the company between 1893 and 1923. The elegant design of the pub is due to the fact it occupies the former ballroom of the former Great Eastern Hotel which was next to the station.

Hand and Flower *1 Hammersmith Road, W14 8XJ*
Dating from the late 18th century, this name references the presence of a plant nursery in the local area at that time.

Hand and Racket *25-27 Wimbledon Hill, SW19 7NE*
As a pub in Wimbledon, it's perhaps no surprise they've chosen a tennis theme for their name, much like the 'Centre Court' shopping centre right by the railway station. Of course, technically the tournament takes place closer to Southfields than here but. as it's called the Wimbledon Championship, can't really fault the people who set this pub up in the mid-90s.

Hand and Shears *1 Middle Street, EC1A 7JA*
This relates back to the presence of a cloth fair close to the pub when it first opened in the 16th century. The fair, held at Bartholomew, was said to be the biggest event for cloth in England at the time. There are various interpretations of what aspect of this fair the name references. Suggestions include the use of shears at the fair, the cloth and the tools they used or even the fact the Lord Mayor of London opened the fair each year by cutting the first piece of cloth himself.

Hand of Glory *240 Amhurst Road, Lower Clapton, E8 2BS*
This pub is themed around English pagan folklore, which explains the name for this pub. A Hand of Glory is the left hand of a hanged man. This was then pickled and dried before being dipped in wax. It was meant to ward off evil spirits.

Hangar *37 The Oval, Sidcup, DA15 9ER*
and 35 Bellegrove Road, DA16 3PB
The name of these two micropubs in south-east London comes from the first name of their founder, Cliff. Put the two together and what do you get?

The Hansom Cab *84-86 Earls Court Road, W8 6EG*
Hansom Cabs were the taxis of their day. Designed by Joseph Hansom in 1834 they were quicker and safer than the original hackney carriages and subsequently became a firm fixture on the streets of London during the Victorian era. They were replaced in the early 20th century when taxis became motorised. A Hansom Cab used to hang from the ceiling here for many years but has since been removed.

The Harlequin *27 Arlington Way, EC1R 1UY*
Located next door to Sadlers Wells Theatre, this pub is named in honour of the clown figures which were featured during the early 19th century heyday of English pantomime. Harlequins originated in Italian comedic theatre, with all the masks, and were the comic servant role. The leading star of the genre at that time, Joseph Grimaldi, was a regular performer at Sadlers Wells and his farewell speech is framed inside the pub. Grimaldi is commemorated on nearby Exmouth market with perhaps the best blue plaque in London, which just reads CLOWN.

The Harlequin.

Harrild and Sons *26 Farringdon Street, EC4A 4AB*
A reference to the first occupiers of the building, the factory of a printing company that operated from this site since the early 19th century until 1949, with the current building dating from 1886. After closure it was converted to offices and only became a pub in February 2014.

Harvest Moon *141-143 High Street, Orpington, BR6 0TW*
Located at the front of the Walnut Shopping Centre, this name reflects the fact the land here was once owned by a Mrs Vinson whose family grew large quantities of soft fruits.

Hawkins Forge *110 Battersea Rise, SW11 1EJ*
This commemorates the former use of the site before it became a pub. When the blacksmith James Hawkins died in the early 19th century, his widow Ann sold the lease to George Finch, landlord of the Half Moon in Putney, in 1809. By the 1840s the Finch's had acquired the freehold and built

The Hercules.

The Heron.

this pub, which briefly was known as the Universe, then the Railway Tavern, and the Dog and Duck and most recently The Duck before it acquired its present identity in 2015.

Hector and Noble *233-235 Victoria Park Road, E9 7HD*
Continuing with the Victoria theme, the park is five minutes' walk from here; this pub is named after the Monarch's two favourite dogs.

The Hemingway *84 Victoria Park Road, E9 7JL*
While this pub is obviously named after the American writer Ernest Hemingway, author of *For Whom The Bell Tolls* and *The Old Man and The Sea*, but we discovered that the owner has named all of his pubs after US writers, including the Hunter S (Thompson).

The Henry Addington *22-28 Mackenzie Walk, Canary Wharf, E14 4PH*
Henry Addington was Prime Minister from 1801-1804 and was at one time a doctor to George III. This pub claims to have Britain's longest pub bar. His ties to the docklands are that he opened them in 1802.

Henry Holland *39 Duke Street, Marylebone, W1U 1LP*
This pub dates from 1958 and was named after an architect from the late 18th century who remodelled the Southill Park Estate in Bedfordshire for Samuel Whitbread, founder of the brewery which owned the pub at that time. He also designed the first iteration of the 'Marine Pavilion' in Brighton, a much more modest and understated affair than the Royal Pavilion which stands today.

Hercules *2 Kennington Road, SE1 7BL*
This pub is a short walk from the site of Astley's (Philip, not Rick) Amphitheatre, with Astley thought of as the father of modern circus. As a result, when a pub was reinstated here in 2019 after an extended period as a Chinese restaurant, Hercules was chosen in homage to the famous circus strongmen of the same name.

Hercules Pillars *18 Great Queen Street, WC2B 5DG*
This pub is located right by London's massive Freemasons Hall. In Masonic Temples there are sometimes two pillars that are passed through before reaching the main meeting hall which have been known as Hercules Pillars, from where the name here derives.

Hermits Cave *28 Camberwell Church Street, SE5 8QU*
Takes its name from Samuel Matthews, a gentleman in the late 18th century who retreated to a cave-dwelling existence in nearby Dulwich woods following the death of his wife. Matthews was murdered in his cave in 1802 in a case that was never solved, having been badly beaten in an earlier incident in 1798.

Hero of Maida *55 Shirland Road, W9 2JD*
Of the six tube stations in London named after pubs, Maida Vale is arguably hardest to guess. The original pub in question, the Maida, was often known interchangeably as the Hero of Maida, opened shortly after John Stewart's victory over the French at Maida (in Italy), in 1806. Stewart was depicted on the early pub signs as the 'Hero of Maida'. The name soon spread to the whole area. This pub is not the original Hero, but it is still welcome to see the return of this historic entry into London folklore.

The Heron *Norfolk Crescent, W2 2DN*
Previously called The Fountain in homage to the water gardens in the adjacent 1960s modernist private estate, the pub was renamed in the last decade after a heron which visited the estate garden's pond began to pop into the pub's own patio area.

High Cross *350 High Road, Tottenham, N17 9HT*
This unique example of a Mock Tudor frontage former public toilet-turned-pub takes its billing from a nearby monument, originally built at the start of the 17th century to mark the centre of Tottenham Village.

Hogarth *58 Broad Street, Teddington, TW11 8QY*
William Hogarth was a hugely influential 18th century painter and political satirist who lived a few miles up the road in Chiswick. Two of his most famous works reference alcohol. Given the positivity expressed in his Beer Alley cartoon, we think he'd be happy his name has been used for a pub. However, we're not so sure he'd be as happy about gin's massive resurgence in recent years, given how he depicted the impact of 'Mother's Ruin' in London in the 18th century in his Gin Lane cartoon.

Hoop and Toy *34 Thurloe Place, Kensington, SW7 2HQ*
Thought to be the oldest pub in Kensington, it began life with the ubiquitous names of the Grapes and then Hoop and Grapes. The pub gained its present name way back in 1784 after the landlord at the

time built a pub in Chelsea called the Bunch of Grapes and altered the name here to avoid confusion. The Toy in the name is thought to refer to a rubber ball given to horses by their owners as they were tethered outside the pub in order to stop them damaging their gums.

The Horatia *98-102 Holloway Road, N7 8JE*
Named after the daughter of Lord Horatio Nelson (see Tapping the Admiral) and Lady Emma Hamilton (see Lady Hamilton), although she never acknowledged Emma as her real mother. She would go on to marry a reverend and have ten children with him, before dying in 1859. She is buried in Pinner.

The Horn of Plenty *36 Globe Road, Bethnal Green, E1 4DU*
This pub goes back to antiquity for its name: the Horn of Plenty was also called Cornucopia (Latin for horn of abundance) and has a number of origin stories. Many Greek and Roman gods are depicted with the horn and possession of it is meant to indicate that you will have everything you desire.

Horniman at Hay's *Unit 26 Hays Galleria, SE1 2HD*
Set in the gorgeous Hay's Galleria which used to previously be Hay's Wharf, at the time London's largest port (see Alexander Hay). In the 1840s, as part of William Cubitt's conversion, warehouses were constructed and one was owned by a tea merchant called Frederick Horniman. Now the pub occupies the site in which his warehouse used to be.

Horse and Guardsman *16-18 Whitehall, SW1A 2DY*
Formerly called the Lord Moon of the Mall, this name comes from its proximity to the Horse Guards building, the historic base of the Guardsman's horses which looked after the Monarch dating back to Charles II's reign.

Horse and Wig *14 Fulwood Place, WC1V 6HZ*
Examining the evidence here, we think it's a fairly open and shut case that the 'wig' in the name is because of the pub's proximity to London's legal heartland. And the horse? The pub is located a short distance from a mews called Jockey's Fields, which dates in 1720 and named after the field which it was built on. Appropriately enough the houses on this mews were originally used as stable accommodation for the houses on the next street, Bedford Row.

House of Hammerton *99 Holloway Road, N7 8LT*
This pub serves as the tap room for the nearby Hammerton Brewery. This is a revival of the original Hammerton Brewery which operated between 1868 and the 1950s. A current member of the Hammerton family decided to resurrect the business in 2014.

Hung, Drawn and Quartered *26-27 Great Tower Street, EC3R 5AQ*
This rather grisly name derives from the fact public executions used to be held in the area immediately in front of the pub; there is a sign outside that quotes a passage from Samuel Pepys diary in October 1660 where he recalls witnessing an execution here in such a fashion. The building only became a pub in the 1990s and was previously a bank; hopefully none of the customers ever felt they'd been hung, drawn and quartered by excessive bank fees...

The Hudson Bay *1-5 Upton Lane, E7 9PA*
Forest Gate's Wetherspoons is named after the Hudson Bay Company which was created by Sir John Pelly, one of the local landowners.

Hunters Moon *86 Fulham Road, SW3 6HR*
It's a multi-layered explanation to this particular lunar-based name. The owner Hubert's thought process began with considering the link between the night's sky and frivolity in human history. From this came a reflection on the name of George Orwell's perfect yet totally fictitious pub, the Moon Under Water. Hubert then recalled one of his favourite pubs in the UK in south Wales had the name Hunters Moon. Building on this, he realised it also linked nicely to his own first name, Hubert, as he was named after St. Hubert, patron saint of hunting and feasting. The cherry on the cake came with the fact the pub opened in October 2019, the month which the Hunter's Moon, marking the coming of winter, appears in the night sky.

Hunter S *194 Southgate Road, N1 3HT*
This pub has the same owner as the Hemingway and carries on the theme of being named after another author who liked to do things to excess. An explanation for the name used to be offered on their website where they stated how the author knew 'good food and drink and the excess makes for the good life', and how they'd decorated their pub in his style, including oversized taxidermy.

Huntsman and Hounds *Ockenden Road, Upminster, RM14 2DN*
Located on an old coaching route, this pub was a meeting point for the local hunt.

The Hydrant *Equitable House, 1 Monument Street, EC3R 8BG*
Fire is the inspiration here, as the pub is opposite the Monument to the Great Fire of London. The pub was even officially opened by the London Fire Brigade's commissioner, Ron Dobson CBE, in 2015.

The Ice Wharf *Suffolk Wharf, 28 Jamestown Road, Camden Town, NW1 7BY*
Camden's branch of Wetherspoons is named after the ice wharf that sat next to the site of the pub. The wharf was opened in 1837 to handle imported ice from Norway.

Inn 1888 *21a Devonshire Street, W1G 6PD*

This was the year the pub first opened as The Devonshire Arms, before being renamed in the last 20 years to reflect its birth date. Some sources have suggested the building dates from 1898 and that 1888 was chosen because it was the year of the Jack the Ripper murders. We asked the pub directly who told the first story, hence our choice!

Inn1888.

The Jackalope.

Inn of Court *18 Holborn, EC1N 2LE*

This is primarily named after the professional associations for barristers in England and Wales, which form the Inns of Court. These four Inns (Grays Inn, Lincolns Inn and Inner and Middle Temple) are all located within the pub's immediate vicinity.

The Iron Duke *11 Avery Row, Mayfair, W1K 4AN*

While the Duke of Wellington has many pubs named after him, the Iron Duke was the nickname he acquired when he was Prime Minister between 1828 and 1830. Failing to see public anger due to famine, high unemployment and the unpopular Corn Laws, he said there was no need for constitutional reform after the opposing Whig Party under Earl Grey called for it. Upon hearing this news that the rioters who had gathered in London headed for Wellington's residence, Apsley House, he responded by barring the windows with iron bars. He would soon resign and the Great Reform Act would come just two years later, in 1832.

Island Queen *87 Noel Road, Islington, N1 8HD*
The Island Queen is named after two identical steamships which cruised up and down the Mississippi and Ohio rivers in the USA. The first one burned in a fire in 1922 and the second in 1947 when the chief engineer accidentally used a welding torch on the fuel tank.

Italian Greyhound *62 Seymour Street, W1H 5BH*
Formerly an Italian restaurant called Bernardi's, this reopened as an Italian pub post-lockdown. The restaurant had this particular dog as its logo and so it was not only retained but used for the pub's name, the owners stating that the breed is elegant and nimble yet approachable, feminine and quintessentially Italian.

Italian Job *13 Devonshire Road, Chiswick, W4 2EU*
and 130 Cadogan Terrace, E9 5HP
Sorry to disappoint fans of the 1969 cult movie (and surely a smaller contingent of fans of the modern remake), but these two pubs are Italian craft beer venues rather than a themed homage to Michael Caine and his mini. You do wonder though after a few birra how many people start doing their best "You were only supposed to blow the bloody doors off!" recital. Maybe this could be a location for Steve Coogan and Rob Brydon in a future series of *The Trip...*

J J Moon's *Various - Five Pubs in London*
These Wetherspoons in London are named after a fictional character called J J Moon. This is also a play on 'Moon Under Water' which is another popular name for some of Spoon's establishments (See Moon Under Water)

Jackalope *43 Weymouth Mews, W1G 7EQ*
The most recent addition to a small pub chain all named after animals, other venues include the Resting Hare (we'll come to that later) and the sadly closed Holborn Whippet. Unlike those two, this animal is entirely fictitious, a mythical creature that resembles a rabbit with antelope horns, hence the name (Jack Rabbit and Antelope merged).

Jack Horner *234-236 Tottenham Court Road, W1T 7QN*
This former bank was converted into a pub in 1994. Its name is a nod to its sub-brand within the Fullers family of pubs, it's one of their 'ale and pie' houses so it was decided to use the name of the boy in the old nursery rhyme who ate his Christmas pie in the corner and was chuffed when he found a plum in it.

Jack and Jill *Longslands Avenue, CR5 2QJ*
Named after the famous nursery rhyme for two reasons: the pub is on a hill and when the site was being cleared for its construction the remains of an old well were discovered. So, Jack and Jill really could have gone up this hill to fetch a pail of water.

Jamaica Wine House *St Michael's Alley, Cornhill, EC3V 9DS*
Sited on the location of London's first coffee house in the mid-17th century, it would become known as the Jamaica Coffee House as the English colonised the island around the same time. Samuel Pepys was known to frequent the establishment. The wine comes in as there is a wine bar downstairs. Despite this confusion, it is very much a pub!

Jamaica Wine House.

Jerusalem Tavern *55 Britton Street, EC1M 5UQ*
This references the fact the local area here was once dominated by a powerful religious grouping called the Priory of St John of Jerusalem. Despite appearances suggesting a real historical vintage, the pub has only been in its current location since 1996 but can trace an interrupted legacy in this area back to the 17th century.

Jobbers Rest *St Mary's Lane, Upminster, RM14 3LT*
The only pub in Britain with this name, this Havering establishment dates back to 1860. The exact meaning behind its name has never been recorded but one potential theory is that it was named after the many casual employees that existed at the time and whom the pub served. Andy Grant, a fount of knowledge on Havering pubs, pointed out to us that in the Victorian era the term jobber would apply to this group of workers and not merely odd-job men who we'd associate with the term today.

The Job Centre *120-122 Deptford High Street, SE8 4NP*
This name caused outrage when it was opened in 2014 as the building used to be the actual job centre. The irony of the name was lost on some who saw it as a blatant sign of gentrification but the anger has now died down.

John Salt *131 Upper Street, Islington, N1 1QP*
John Salt is one of the pioneers of the photorealist art movement which developed in the 1960s. Many of his works are of American cars after he moved there but he returned to England in 1978. He lives in Shropshire where he continues to paint.

John Snow *39 Broadwick Street, Soho, W1F 9QJ*
John Snow is the father of epidemiology: in 1854 he worked out that a cholera outbreak in Soho came from a particular water pump on the street where the pub is now located and convinced the authorities to close off the pump. This stopped the outbreak, disproving the existing theory that cholera was spread through the air but in fact through the water supply. He also led the way in anaesthetics, famously giving chloroform to Queen Victoria while she was giving birth.

John the Unicorn *157-159 Rye Lane, SE15 4TL*
This pub in Peckham has one of the most heart-warming stories in this whole book. The original landlord's daughter had a fluffy unicorn called John which she was obsessed with, so he decided to name the pub in honour of her favourite toy. How lovely!

Joiner on Worship *2 Paul Street, EC2A 4JH*
A reference to the original use for this building as a furniture warehouse and subsequent showroom for the furniture company Moore and Hunton, which today is a massive three floor pub! Worship refers to one of the roads it spans as opposed to devotion to the furniture trade.

Jolly Butchers *168 Baker Street, Enfield, EN1 3JS*
and 204 Stoke Newington High Street, N16 7HU
Many pubs in London feature 'Jolly' at the start of their names. Supposedly this is because the word conjures up images of good times and feeling happy. They are often followed by an occupation, possibly because like most pubs which feature 'arms' at the end, it could be a sign of the former landlord's day job (we've used Butchers as an illustration, but there are many others).

Joyce *294 Brockley Road, SE4 9RA*
The owner named this after his grandmother, who lived on a nearby street here in Brockley.

Jugged Hare *172 Vauxhall Bridge Road, SW1V 1DX*
Part of the Fullers 'Ale and Pie' group, this pub is named after the classic British pie dish.

Kentish Belle *8 Pickford Lane, DA7 4QW*
Located barely a double arrow away from Bexleyheath station, this micropub borrows its title from a named train operated by British Railways between London Victoria and the Kent Coast from 1951

to 1958. The pub hit the headlines when it opened on the stroke of midnight on Monday 12 April 2021, the moment pubs were able to reopen (only for outdoor drinks at that point) following the 2021 national lockdown.

Kentish Drovers *71-79 Peckham High Street, SE15 5RS*
Takes its name from the cattle drovers, a drover being someone who walks livestock to market, passing through Peckham on the way to Smithfield Market. Sheep drovers also played their part in naming another south London pub, sadly currently closed, 'Sun in the Sands', after the sight of the setting sun being surrounded by dust kicked up by the sheep on their way into London.

The King & Tinker *Whitewebbs Lane, Enfield EN2 9HN*
King James I went hunting in Enfield before being separated from his entourage. He decided to grab a pint at this very pub and struck up a conversation with a tinker. Unaware of who he was talking to, the tinker asked who he was. James replied saying it would only be revealed when everyone else was hatless, leaving the tinker puzzled. When the King's men caught up with him, they immediately removed their hats and the tinker realised who his new drinking buddy was.

The King of Prussia *363 Regents Park Road, N3 1DH*
This pub in Finchley Central is on the site of a pub with the same name. Prussia was a state in Germany whose expansion under Chancellor Otto Von Bismarck helped to unify Germany in 1871. They had a king from 1701-1918 at the end of the First World War. The old pub was sadly demolished to make way for the office block which currently occupies the site in 1965. The pub was only reopened in 2019 after previously being a restaurant and the owner decided to pay homage by returning it to its original name.

The King's Ford *250-252 Chingford Mount Road, E4 8JL*
There was a ford on the River Lea in this area, which was apparently crossed by King Alfred the Great, leading to it being called King's Ford. This was corrupted over time to become Chingford, allegedly.

The King's Stores *14 Widegate Street, E1 7HP*
This pub is built on the site of a huge army munitions depot that existed during the reign of Henry VIII.

Kiss Me Hardy *Unit 5, Priory Retail Park, 131 High Street, SW19 2PP*
Lord Nelson (see also Tapping the Admiral) bought a grand house in this area in the early 1800s called Merton Place, which is now occupied by Reform Place, and the Nelson and Merton connection could be the explanation behind this pub being named after some of his final words. At the Battle of Trafalgar in 1805, Nelson was famously wounded by a musket shot from the French ship Redoubtable because Nelson refused

The Kings Stores.

to take off his medals, which gave away his importance. As he lay dying, he asked Captain Thomas Hardy, in charge of his flagship HMS Victory, to kiss him, which Hardy did on his cheek. Nelson then said: "Now I am satisfied, thank God I have done my duty" before passing away.

The Knights Templar *95 Chancery Lane, WC2A 1DT*
Named after the 12th century order of knights who owned the land upon which the pub would be built on Chancery Lane. They were influential during the crusades and as well as being fighters, they were also instrumental in finance and fortifications. They were disbanded in 1312 after pressure from King Philip IV of France who had a huge debt with the knights which was subsequently cleared.

Knowles of Norwood *294-296 Norwood Road, Norwood, SE27 9AF*
This south London tavern is so named because the site used to be a hardware store called Knowles before it was converted into a pub.

Lady Hamilton *289-291 Kentish Town Road, NW5 2JS*
Lady Emma Hamilton was a famous actress who married Sir William Hamilton, Ambassador to Naples, in 1791. She soon met Admiral Horatio Nelson and the two began having an affair which made her a celebrity in her own right. They had a daughter together (see The Horatia) and were very much happy despite the scandal. Upon Nelson's death at the Battle of Trafalgar in 1805, she was devastated. Things only got worse when she wasn't allowed at his funeral and was cut out of the inheritance, despite Nelson writing her into his will. Cut off from society, she moved to France where she died in 1815.

The Lady Ottoline *11A Northington Street, WC1N 2JF*
Located in Bloomsbury, this pub honours Lady Ottoline Morrell who was a socialite in the early 20th century and lived nearby. She was part of the Bloomsbury Set, hosting the likes of Virginia Woolf, W B Yeats (see W B Yeats), T S Eliot and D H Lawrence (who used her as inspiration for *Lady Chatterley's Lover*, apparently). She also had a long affair with Bertrand Russell.

The Lady Ottoline.

The Lass O'Richmond Hill.

Lamb & Flag *33 Rose Street, Covent Garden, WC2E 9EB and 24 James Street, W1U 1EL*
The image of a holy lamb bearing a St George's Flag was previously a symbol for the Knights Templar who have a rich history in the Middle Ages and were influential in medieval London (See Knights Templar).

The Lamb and Trotter *6 Little Britain, EC1A 7BX*
Being located just to the south of Smithfield Market, you could be forgiven for assuming this name links to the animals traded there. Instead, it references two feuding families who lived on Little Britain and whose run-ins were chronicled by the writer Washington Irving in 1820.

Landseer Arms *37 Landseer Road, Upper Holloway, N19 4JU*
Edwin Landseer was a Victorian artist and is remembered in this Islington pub. His most famous works are a painting called *The Monarch of the Glen* depicting a stag and he also sculpted the lion statues which adorn Nelson's Column on Trafalgar Square.

Lass O'Richmond Hill *8 Queens Road, TW10 6JJ*
Refers to an 18th-century folk song, the 'Lass O'Richmond Hill';
however, the Richmond in question is in Yorkshire rather than beside
the Thames. London's Richmond actually takes its name from its
northern compatriot after Henry VII built a palace here and named
it after his favourite hunting grounds up north, as he was Earl of
Richmond. The pub is gamely carrying on that regal tradition of
borrowing from its Yorkshire namesake.

Latymer's *157 Hammersmith Road, W6 8BS*
This early 1980s pub is named after the independent school which
is based to the west of Hammersmith and also lent its name to the
impressive 1930s apartment block opposite the pub.

Lemon Tree *4 Bedfordbury, WC2N 4BP*
The name here is thought to be a nod to the citrus fruit available at
nearby Covent Garden during its long tenure as a market, long before its
repurposing as a bustling area of bars, restaurants and human statues...

Leather Exchange *15 Leathermarket Street, SE1 3HN*
Bermondsey was one of the epicentres of the UK leather industry,
being nicknamed 'the land of leather' and estimated to be producing
one third of the country's leather in the late 18th century. The pub here
was the tavern that served the leather market based in a separate part
of the building which opened at the end of 1879. The Exchange itself
only had a brief spell in its original function before becoming offices
for companies involved in the leather trade by the early Edwardian
period.

The Ledger Building *West India Quay, 4 Hertsmere Road, E14 4AL*
Being located in the docklands, this Wetherspoons is found in the
building that used to house the ledgers detailing all of the goods that
come through the docks.

The Lexington *96-98 Pentonville Road, Islington, N1 9JB*
Many people believe this pub is named after Lexington, Kentucky due
to the fact it has over 150 bourbons on offer. However, it's actually
named after Lexington Avenue in New York City, which was famously
mentioned in the Lou Reed song 'Waiting for the Man'.

Leyton Technical *265B High Road, Leyton, E10 5QN*
This building began life in 1896 as a technical college, although then
spent most of its pre-pub life as Leyton Town Hall. However, calling it
the Town Hall may have caused confusion with people turning up to
object to planning applications only to find themselves in a pub, so we
can see why they plumped for the Technical.

Libertine *125 Great Suffolk Street, SE1 1PQ*
The owner of this pub chose this name in homage to his own support for the rebellious types in history and those who have been considered Libertines, which is defined as people who value the importance of pleasure through the physical senses, and the walls are decorated with various 'libertines' of history such as Casanova.

Liberty Bell *Mercury Gardens, Romford, RM1 3EN*
Now we'd love to tell you that this heralds a symbolic link between an icon of American independence and Romford but I'm afraid that's not the story here! The Liberty being referred to is the old name for the area, when Havering was a Royal Liberty owned by the Crown by virtue of a royal residence at Havering Palace. The 'liberty' continued even after the Palace was sold into private hands in 1828 before finally vanishing in 1892, as the area was incorporated into Essex.

Liberty Bounds *15 Trinity Square, EC3N 4AA*
This pub stands just beyond the City of London's boundaries, or its liberty, as it was more commonly referred to in days gone by.

The Lillie Langtry *19 Lillie Road, West Brompton, SW6 1UE*
Named after the Jersey born actress and socialite who was the Nell Gwynne of her day in the 1870s onwards. She would have affairs with the Prince of Wales (later Edward VII) and would also become a thoroughbred horse trainer.

The Lion & Unicorn *42-44 Gaisford Street, NW5 2ED*
The Lion & Unicorn is the symbol of the United Kingdom, with the lion representing England and the Unicorn representing Scotland. Their inclusion on the royal coat of arms dates back to 1603 when James VI of Scotland also became James I of England.

Little Driver *125 Bow Road, E3 2AN*
The name here is thought to have evolved over time, potentially starting out as the Little Drover on account of being on one of the drovers' routes into London. The suggestion is it changed to Driver when Bow Road railway station opened two doors along from the pub in 1892. The station closed in 1949 and its building is now used by a bookmakers'.

Little Green Dragon Ale House *928 Green Lanes, N21 2AD*
Enfield's first Micropub harks back to the former pub which used to be nearby which was called The Green Dragon. Given it's a micropub, it's a nice continuation of the name, albeit with 'little' affixed to the start. The dragon is, in fact, the one alleged to have fought St George, which is often depicted as being green (See George & Dragon).

Little Driver.

Little Windsor *13 Greyhound Road, Sutton, SM1 4BY*
This name is a charming case of understated modesty. Previously called the Windsor Castle, the present landlord changed the name soon after he took over in 1998 to avoid confusion with another Windsor Castle pub in nearby Carshalton and to reflect the fact this pub is on the small side.

Lockhouse *3 Merchant Square, W2 1JZ*
A reference to the fact this modern pub sits opposite Paddington Basin. It's on the ground floor of the Merchant Square development, one of the many new buildings which have sprung up here in recent years as part of the redevelopment of the area.

London Apprentice *62 Church Street, Isleworth, TW7 6BG*
Two potential theories exist for this uniquely named West London pub, one being the fact that apprentices from the livery companies in the city would row out here in their spare time, the other being that it is named after the old song 'The honour of the Apprentice of London'. What isn't in any doubt, though, is the fact the pub once saw the visit of one of the world's biggest media tycoons, Rupert Murdoch. The Australian mogul dropped in here in February 1989 to show his face at the staff party to mark Sky Television taking to the air, which was, and still is, based at studios nearby in Osterley.

London Hospital Tavern *176 Whitechapel Road, E1 1BJ*
One of only a select few pubs in the UK with Hospital in their name, this establishment is based right next door to the Royal London Hospital which also indirectly has given its name to an earlier pub in this book, the Good Samaritan, by virtue of its historic logo.

London and Rye *109 Rushey Green, SE6 4AF*
This Wetherspoons takes its name from the fact it is located on the old coaching route between the capital and the charming East Sussex town of Rye.

London and South Western *276-288 Lavender Hill, SW11 1LJ*
The pub is named after the first railway company to put a line through the area in 1838, running between Nine Elms and Woking. Clapham Junction only arrived in 1863, a joint venture between the LSWR and the two other companies who had lines running through the area, the London, Brighton and South Coast Railway and the West London Extension Railway.

Long Haul *149 Long Lane, Bexley, DA7 5AE*
After having their original opening date put back due to the small matter of the COVID-19 pandemic, the owners of this Bexley pub thought Long Haul neatly summed up how they felt getting this venture over the line, while also linking back to their location on Long Lane.

Long Pond *110 Westmount Road, Eltham, SE9 1UT*
Named after a pond of the same name in the nearby Eltham Park North.

Loose Box *51 Horseferry Road, SW1P 2AA*
Linked to the animal in the street name the pub is on, a Loose Box is a compartment in a stable big enough for a horse to move around in.

Lord Aberconwy *72 Old Broad Street, EC2M 1QT*
Located next to Liverpool St station, this pub remembers the last chairman of the Metropolitan Railway, Lord Aberconwy. He served from 1904-1933 when the railway was amalgamated with the other companies' lines to form the London Passenger Transport Board.

Lord Clyde *27 Clennam Street, Southwark, SE1 1ER*
This Grade-II listed pub remembers Sir Colin Campbell, the famous 19th Century army officer, albeit via his hereditary title. There is another pub bearing his name in Kilburn (See Sir Colin Campbell).

Lord Denman *270-272 Heathway, Dagenham, RM10 8QS*
Thomas Denman was Lord Chief Justice between 1832-1850 and for a time lived nearby to the pub's location. He is best remembered for presiding over Stockdale v Hansard which led to parliamentary privilege being upheld.

Lord John Russell.

Lord Hill *40 Watling Street, Bexleyheath, DA6 7QG*
Thought to be named after the Viscount Rowland Hill, a leading General for Wellington in the Napoleonic Wars, the pub opened in 1824, a mere nine years after the battle of Waterloo.

Lord John Russell *91-93 Marchmont Street, WC1N 1AL*
Lord John Russell was twice Prime Minister, firstly between 1846-52 and a shorter spell from 1865-66. Although the pub is near Russell Square there is no obvious connection as Russell Square was built in the early 1800s and is named after the family name of the Dukes of Bedford who own the land. However, a chat with the landlord revealed it was named in 1909 after a commoner, John Russell, bought the pub and looked for another famous John Russell to name the pub after. Turns out there was one who was PM and was also the Earl of Bloomsbury, a remarkable coincidence given the pub is in Bloomsbury.

Lord Kitchener *49 East Barnet Road, New Barnet, EN4 8RN*
Herbert Kitchener became the Secretary of State for War in 1914, at the start of the First World War. His portrait is famous for being on the army recruitment poster (where he is pictured pointing at the reader saying he wants you) which was instrumental in signing up soldiers at the start of the war.

Lord Northbrook *116 Burnt Ash Road, SE12 8PU*
This pub in Lee, near Lewisham, is named after the Liberal politician who served as Chancellor between 1839-41. He later would serve as First Lord of the Admiralty under Lord John Russell.

Lord Palmerston *252-254 Forest Road, Walthamstow, E17 6JG and 33 Dartmouth Park Hill, NW5 1HU*

London has two pubs that remember Lord Palmerston, who dominated UK politics between the 1830s and the 1860s. He was Foreign Secretary three times and Prime Minister twice, and despite being unpopular with the political establishment and Queen Victoria his popularity with the public helped keep him at the heart of Westminster. He died in 1865, a few months after increasing his majority in Parliament, and to date is the last PM to die in office. He also received a state funeral, becoming only the fifth non-royal to be allowed one.

Lord Palmerston.

Lucky Rover.

Lord Raglan *Various - Three Pubs in London*

These pubs in London remember Lord Raglan, who was in the British Army during the 19th century. He served as Wellington's Military Secretary in the Napoleonic Wars when he was a junior officer. He would then go on to work through the ranks so that by the time of the Crimean War, he commanded the British troops as a general. He would be responsible for the infamous Charge of the Light Brigade and his failures of leadership were panned in the press. He died suddenly during the Siege of Sevastopol in 1855.

The Lord Tredegar *50 Lichfield Road, E3 5AL*

Godfrey Morgan, 1st Viscount Tredegar, is literally honoured at this East End pub. He is famous for being a captain in the 17th Lancers which famously participated in The Charge of the Light Brigade. Tredegar survived, along with his horse, Sir Briggs.

Lord Wargrave *40-42 Brendon Street, W1H 5HE*
Edward Goudling was a Conservative politician at the start of the 20th century. He would be ennobled in 1922 to become Baron Wargrave. Outside of politics he was Chairman of Rolls-Royce.

Lord's Tavern *Lord's Cricket Ground, St John's Wood Road, NW8 8QN*
Thomas Lord was the founder of the Marylebone Cricket Club, which he established in 1787. In 1814 the club moved to its present site in St John's Wood and, when there was an increasing need to provide more refreshments to spectators, the tavern was built initially in the ground before being moved beside the Grace Gates. The pub is something of a pilgrimage for cricket fans worldwide.

Lucky Rover *312 Hook Road, KT9 1NY*
This pub has always been known by this distinctive name and is the only one in the country. No formal explanation has ever been recorded but it is thought to either be named after a boat, as featured on the pub sign, or a reflection of the first landlord's wandering spirit! On a complete tangent, the pub is also used as the registered headquarters for the Monster Raving Loony Party at election times!

Lyceum Tavern *354 Strand, WC2R 0HS*
The pub is built on the original site of the Lyceum Tavern, which burnt down in 1830 and was rebuilt slightly to the north of where the pub now stands.

Lyttelton Arms.

Lyric *37 Great Windmill Street, W1D 7LT*
Renamed in 1890 after the nearby West End theatre which opened two years earlier, the pub had previously been called the Ham and Windmill. That original name isn't as exciting as it sounds, it merely came from two establishments merging.

Lyttelton Arms
1 Camden High Street, NW1 7JE
We're very pleased to say we do have a clue about the story behind this name. The pub was renamed in 2014 in

Mabel's Tavern.

honour of the late Humphrey Lyttelton, chairman of the long-running Radio 4 comedy panel show *I'm Sorry I Haven't A Clue* which helped plant the tube station opposite into the public consciousness with their nonsensical game, Mornington Crescent. In the 1990s Lyttelton and the show's other panellists helped keep the station on the tube map after it shut indefinitely when lift repairs were required; at one stage a permanent closure was a real possibility as an alternative for having to invest in the new lifts. Lyttelton and his colleagues attended the reopening ceremony in 1998; a plaque that marks their fundraising efforts can be found within the station.

Mabel's Tavern *9 Mabledon Place, WC1H 8AZ*
Mabel Macinelly was born in Ireland before marrying to become Mabel Hamilton, who owned Hamilton House on the same street where this pub is found. She died in the 1970s but she has since 'revisited' her pub by haunting it on numerous occasions, including operating the dumb waiter (the lift that carries food from the kitchen) in the early hours of the morning.

Macbeth *70 Hoxton Street, N1 6LP*
Known for much of its life as The White Horse, it was renamed in 1978 after one of Shakespeare's most famous, and cursed if you believe tradition, plays. The name comes from the fact one of the pub's interior walls is decorated with a mural of a scene from the play. The reasons behind why this was first installed are unknown, but Shakespeare himself was thought to live in Hoxton at one point.

The Mad Hatter.

Magpie & Stump.

Mad Bishop and Bear *1st Floor, Paddington Station, W2 1HB*
This station pub takes its curious name from one figure crucial to the development of the station here and another who has kept it in the public psyche! The Mad Bishop refers to the Bishop of London, who owned the land the station was built on before selling it to the railway company at what was considered a knock-down price, hence the 'Mad'. The 'Bear' bit is slightly easier to guess, being everyone's favourite Marmalade lover from Peru, Paddington.

Mad Hatter *3-7 Samford Street, SE1 9NY*
This building was originally home to a hat factory. Sorry to disappoint anyone hoping for something *Alice in Wonderland* related but there will be something on that theme later…

Magdala Tavern *2A South Hill Park, NW3 2SB*
The pub opened in the late 1860s and took its name from a British military victory in 1868 at the battle of Magdala, a fishing village in what was then Abyssinia. The pub gained infamy in the 1950s as the location where Ruth Ellis, the last woman to be hanged in Britain, shot her lover David Blakely in 1955. When the pub closed in 2014 there were fears it would be converted into housing in this desirable part of London but, thankfully, in May 2021 the Magdala opened its doors once more.

Magpie and Stump *18 Old Bailey, EC4M 7EP*
While the current pub building is relatively recent, there has been a Magpie and Stump on this site for hundreds of years. The curious name is thought to derive its pub sign from when it was merely the Magpie, which showed our feathered friends perched on a stump. In one of its early incarnations it was also used as a spot to watch public executions at Newgate Prison, where the Old Bailey now stands.

Man of Aran *424-426 Alexandra Avenue, Rayners Lane, HA2 9TW*
This Irish pub is named after the 1934 film that depicted life on the Aran Islands, found off the west coast of Ireland.

The Man on the Moon *112 Headley Drive, Croydon, CR0 0QF*
This pub first opened in the early 1960s known as Man in the Moon, in keeping with that old fable. After Neil Armstrong and co made that one huge leap for mankind in 1969, the enterprising landlord here took one small step of altering a single letter of the pub's name to mark that momentous achievement.

Maple Leaf *41 Maiden Lane, WC2E 7LJ*
This pub is found between Covent Garden and The Strand. It's actually a Canadian themed pub, which gives the game away as the maple leaf is the national symbol of Canada.

Market Porter *9 Stoney Street, SE1 9AA*
A double meaning here, with a porter both being someone working at the market (in this case Borough) as well as being a type of strong dark beer. The pub has also become famous worldwide for being used in one of the Harry Potter films as the 'Third Hand Book Emporium'.

Marksman *254 Hackney Road, E2 7SJ*
Located on the corner of Horatio Road and a short distance from a pub called The Nelson, this is thought to be named after the marksman that fired the killer blow at Admiral Nelson during the Battle of Trafalgar in 1805. One of the potential reasons for this is because the pub is located roughly the same distance from the Nelson pub as the actual marksman was from the man himself on that fateful day.

Marquess of Anglesey *39 Bow Street, WC2E 7AU*
This pub was renamed in 1815 after Henry Paget, who was second in command to Wellington (see The Iron Duke) at the Battle of Waterloo. He was made Marquess of Anglesey following the battle, where he lost his leg. When Paget said: "By God, sir, I've lost my leg!". Wellington responded: "By God, sir, so you have!".

The Masque Haunt *168-172 Old Street, EC1V 9BP*
A masque haunt was a type of social event in the reign of Elizabeth I. It would often feature music, drama and dancing which would be performed by masked performers. The rather brilliantly titled 'Office of the Master of the Revels' was located near to the pub.

The Mayflower *117 Rotherhithe Street, Rotherhithe, SE16 4NF*
Named after the famous ship which took the first pilgrims across the Atlantic Ocean to America in 1620. Rumour has it that the planks of the ship were used in the rebuilding of the pub after it returned to London. One thing is for certain, you can legally buy both UK and US postage stamps behind the bar. A pint of lager and 12 first class stamps please!

Mere Scribbler *426 Streatham High Road, SW16 3PX*
This was the nickname of Frances Burney, an 18th century novelist, diarist and playwright who lived in Streatham between 1779 and 1783.

The Metropolitan Bar *7 Station Approach, Marylebone Road, NW1 5L*
Located above Baker Street station, this pub is named after the Metropolitan Line which was the world's first underground railway when it was opened in 1863 between Paddington and Farringdon. The original company was headquartered in the same building as the pub.

The Miller *96 Snowsfield, SE1 3SS*
Originally known as the Miller of Mansfield after the poem by Bishop Thomas Percy about how King Henry II had asked the miller to put his son up for the night. The miller did just that but as rumours were being spread that he'd also been helping himself to some of the King's deer in Sherwood forest, he was somewhat worried for his fate when Henry returned for his son. However, the King took no notice of that and made the Miller a knight! The connection between the poem and this part of London is that the author's father, Anthony Percy, was both a landowner and merchant in Southwark, so Bishop Percy will also have spent his early years in these parts.

The Miller's Well *419-421 Barking Road, E6 2JX*
East Ham's Wetherspoons is named after a supposedly healing well that used to be located nearby when the East End was open countryside back in the early nineteenth century. Sadly, there is no trace of the well left after the area was built up.

Moby Dick *6 Russell Place, Greenland Dock, SE16 7PL*
This pub surfaced in the 1980s as part of a massive redevelopment of the Greenland Dock area which included new residential luxury property developments; the use of the name *Moby Dick* hints back to Greenland Dock's early use in the 19th century by the whaling industry.

Monkey Puzzle *30 Southwick Street, Paddington, W2 1JQ*
When this pub first opened at the end of the 1960s, the owners planted this distinctive type of evergreen tree with its spiky and triangle-shaped leaves in the pub's front garden. A tree is still going strong there today, although the present staff told us that it is not the original tree. The pub acquired some very unexpected international notoriety in 2019 when the Sudanese government accused anti-government plotters of meeting here to discuss overthrowing the regime...

The Montagu Pyke *105-107 Charing Cross Road, Soho, WC2H 0DT*
Montagu Pyke was a prolific cinema builder in the early 20th Century. The site of the pub is where the final one of his 16 cinemas was opened in 1911. However, in 1915 a fire of waste film reels in the basement led to the death of an employee and Pyke was subsequently prosecuted for

manslaughter. The pub was also formerly the Marquee Club which had
moved in 1988. During its history, the club saw a huge range of bands
perform there between the 1950s and the 1990s.

The Myddleton Arms.

The Monkey Puzzle.

The Moon and Sixpence *250 Uxbridge Rd, Pinner, HA5 4HS*
This Wetherspoons used to be a branch of Barclays, hence the monetary
link in the name. It is also the title of a novel by W Somerset Maugham
published in 1919, where a stockbroker gives up his family and career
to become an artist.

The Moon Under Water *Various - Five Pubs in London*
Many pubs across the country are named the Moon Under Water, most
of which are Wetherspoons. The name comes from a 1946 essay by
George Orwell for the Evening Standard in which he describes his
ideal pub.

Morpeth Arms *58 Millbank, Westminster, SW1P 4RW*
Named in honour of Viscount Morpeth, the man responsible for the
development of the area which is today Pimlico, the pub first opened
in 1845.

The Mossy Well *The Village, 258 Muswell Hill Broadway, N10 3SH*
Muswell Hill's branch of Wetherspoons is named after the medieval
name for the area. The Mossy Well became legendary after a Scottish
king was cured after drinking its water. The pub spent most of the 20th
century as a tearoom before being converted to a pub in the 1980s.

Mother Red Cap *665 Holloway Road, Upper Holloway, N19 5SE*
An Alewife is an old term to cover any brewer who was a woman. It
is believed that they wore red caps which led to them picking up the
nickname of Mother Red Cap.

Mr Fogg's Tavern *58 St Martin's Lane, Charing Cross, WC2N 4EX*
Phileas Fogg is the main character in Jules Verne's famous novel *Around the World in Eighty Days* which was published in 1872.

Mudlark *Montague Close, SE1 9DA*
This pub, located close to the Thames, is named after the description for a person who looks for items of value in the mud of a riverbank. It was particularly popular in London during the 18th and 19th century but there are still plenty of Mudlarks today, as Lara Maiklem's excellent book from 2019 can attest. This part of the river is particularly fertile ground for the larks, in part due to all the entertainment like cockfighting and bear-baiting around the Bankside area of the river when this kind of thing was banned within the old City of London boundaries. Of course, this meant there were plenty of pubs here too! The mud has helped preserve the various items like clay pipes and old coins discarded hundreds of years ago.

Mulberry Bush *89 Upper Ground, SE1 9PP*
Named after the famous nursery rhyme, 'Here we go round the Mulberry Bush' which itself is derived from a children's game in the mid-19th century.

Museum Tavern *49 Great Russell Street, Holborn, WC1B 3BA*
Named after the British Museum, which can be found opposite.

Music Box *Bourne Avenue, Hayes, UB3 1QT*
This is tuned into the presence nearby of the former EMI vinyl factory. In its heyday in the post-war period it was pressing records for huge names like Queen, Pink Floyd and the Beatles, the roll of honour including Britain's biggest selling studio album, *Sgt Pepper's Lonely Hearts Club Band*.

Myddleton Arms *52 Canonbury Road, N1 2HS*
The closest pub to the remaining section of the 'New River' in north London is fittingly named after Sir Hugh Myddleton. Sir Hugh was a wealthy entrepreneur and driving force behind the construction of this artificial river to bring clean water from Hertfordshire into the heart of London in the early 17th century. Myddleton is also commemorated at nearby Islington Green with a 'talking statue': if you scan a QR code on the statue's plinth, you can hear more about Sir Hugh.

Myllet Arms *Western Avenue, Perivale, UB6 8TE*
Named after the Myllet family who were the lords of the manor here back in the 15th and 16th centuries.

Naturalist *14 Woodberry Down, N4 2GB*
This modern pub opened in 2018 and takes its name from the fact it's located right beside Woodberry Down nature reserve.

Navigation *4 Wharf Road, EN3 4XX*
Based right beside the River Lee Navigation, a section of the river which has been channelled into a canal.

The Nell Gwynne Tavern.

Oak & Pastor.

The Nell Gwynne Tavern *2 Bull Inn Court, Charing Cross, WC2R 0NP*
Tucked away down an alley, not far from Covent Garden, is the Nell Gwynne. Nell, born and raised nearby in St Martin in the Fields, sold fruit at Covent Garden before gaining fame as an actress on the Drury Lane stage in the 17th century, and was praised by Samuel Pepys in his diaries.

The Nellie Dean of Soho *89 Dean Street, Soho, W1D 3SU*
'Nellie Dean' is in fact a song that was written in the early 1900s and became the signature song of the music hall entertainer Gertie Gitana, who is also remembered at this Soho pub. The song was sung in pubs across the land and would later become a musical.

New Cross Turnpike *55 Bellegrove Road, Welling, DA16 3PB*
This pub is built on the old road which used to link Dover and London and during the 18th and 19th century was maintained by the New Cross Turnpike Trust who charged a toll for its use.

The Newman Arms *23 Rathbone Street, W1T 1NG*
Takes its name from Newman Hall, the country seat of the local landowner William Berners who developed the Fitzrovia area during the 18th century.

Ninth Life *167 Rushey Green, Catford, SE6 4BD*
This reflects both the pub's location (putting the cat in Catford...) as well as the ethos of this modern venue which also hosts immersive theatre shows and art exhibitions.

Ninth Ward *99-101 Farringdon Road, EC1R 3BN*
Named after a district in New Orleans, this pub being themed in homage to the 'Big Easy' and opened in 2014, one year after the New York venue of this name opened.

The Nightingale *53 Carshalton Road, SM1 4LG*
For many years this pub was called Jenny Lind, after the Swedish singer who took the UK by storm in the 1840s and 50s. A nearby street in Sutton was named after her. Jenny was known as the 'Swedish Nightingale' and her presence lives on here.

Nonsuch Inn *552-556 London Road, Cheam, SM3 9AA*
Takes its name from a nearby park which in turn took its name from a particularly grand Palace of Henry VIII which he had built to eclipse the French King's Palace at Fontainebleau, where the very name itself was chosen to express that no other Palace could hope to match it.

Oak & Pastor *86 Junction Road, Archway, N19 5QZ*
This Archway pub was previously called the Drum & Monkey but now takes its name from the large amount of oak furniture that was saved from an old church (including a full confession booth). The landlady is also Irish, hence the use of Pastor which is an alternative word for Priest on the Emerald Isle.

Observatory *O2 Arena, 31 Entertainment Avenue, SE10 0DY*
Located within the O2 Arena in North Greenwich, this links back to one of the most prominent buildings within Greenwich itself, the Royal Observatory. The O2 doesn't have an observatory: if there was you'd probably just end up seeing people climbing on the roof of the building, an experience people have been able to enjoy since 2017.

Occasional Half *67-77 Green Lanes, Palmers Green, N13 4TD*
Opened in 1995 and formed part of a small collective of pubs, all with half in the title, others including the Significant Half in Clapham Junction, the Better Half in Ealing and the Half and Half in Streatham. The Occasional is the last 'Half' standing!

The Oiler Bar *Royal Victoria Docks, Royal Victoria Beach, E16 1AG*
Another bar on a boat; however, this one was previously a Royal Navy refuelling vessel, hence the name of this pub.

The Old Bank of England *194 Fleet Street, EC4A 2LT*
This rather grand and elegant pub is located in what used to be the
Bank of England's Law Courts. This was from the 1880s until the 1970s
before being converted to a pub in 1994.

The Old Bank of
England.

The Old Bell.

Old Bell *95 Fleet Street, EC4Y 1DH*
Began life as the Golden Bell and later the Twelve Bells. All these
names are probably a reference to St Bride's Church, located directly
behind this building. It's very apt that the name pays homage to the
church, as the pub was built by none other than Sir Christopher Wren
himself, for the stonemasons working for him on the reconstruction of
the church following the Great Fire of London.

The Old Blue Last *38 Great Eastern Street, EC2A 3ES*
Apparently the first pub to sell Porter beer, this tavern is now a famous
music venue having previously been an illegal brothel in the 1990s. A
'Last' is a wooden mould used in shoemaking and the old blue refers
to one of the patterns which would have been used.

Old China Hand *8 Tysoe Street, EC1R 4RQ*
Referencing the background of the pub's current owner and family, an
Old China Hand refers to someone who was born and grew up in China
but has now moved away and is part of an ex-pat community.

Old Dispensary *325 Camberwell New Road, SE5 0TF*
An easy diagnosis here, these premises began life as a medical dispensary set up for poor people in the late Victorian era, a time long before any state health provision. The site was still a chemist's up until the turn of the millennium, having a spell as a restaurant before becoming a pub in the early 2000s.

Old Doctor Butler's Head *2 Mason's Avenue, EC2V 5BT*
Dr William Butler was the court physician to James I. In addition to this lofty position, he also reportedly devised his very own medicinal drink, Dr Butler's purging ale, which sounds a mouth-watering prospect. Some of Dr B's other supposed remedies, such as curing epilepsy by giving the patient a shock like firing a gun right by them or throwing plague sufferers into a cold pond, can certainly be filed in the questionable category. We won't argue with the other pub books we've read which label the good doctor as a quack!

Old Fountain *3 Baldwin Street, EC1V 9NU*
On the land where the pub now stands there was once a large pond fed by an ancient spring. It was first known as the perilous pool, as many poor souls lost their lives there while swimming in it. William Kemp, who acquired the land in the mid-18th century, renamed it as the peerless pool, having developed the old pond into a visitor attraction and London's first outdoor swimming pool, with a separate section for fishing. Kemp sold up in 1805 to Joseph Watts who started draining parts of the pond to build housing and the plug was finally pulled on the peerless pool itself in 1850.

The Old Dr Butler's Head.

Old Frizzle *74-78 Old Broadway, SW19 1RQ*
Named after the elaborate decoration on the Ace of Spades in playing card decks during the 18th and 19th centuries to show the relevant tax had been paid. At the time the description frizzle or frizzled meant something that was highly decorative or ornate. These days being frizzled sounds like you might have been in the pub too long.

Old Hat *128 Broadway, Ealing, W13 0SY*
This pub was once a location for pigeon shooting at the turn of the 19th century, when Ealing was still very rural in nature, and Old Hats was the name of pigeon shooting clubs. The reason for this? Before the pigeons were released and the shooting began they were placed in small holes in the ground which were covered by old hats. When the time came for the contest to begin, the hats were removed with a piece of string.

Old Ivy House *166 Goswell Road, EC1V 7EB*
The sign here shows an old house, covered in ivy, which is thought to have stood on the spot the pub now stands today. There is a very similarly named pub in Peckham, called The Ivy House. While we don't know the tale behind its name(it began life as the Newlands Tavern), it still has a story worth telling as the first pub in London and indeed the UK to be listed as an Asset of Community Value. This helped save it from being turned into housing in 2012 and it reopened as a community-run pub a year later, again being the first of its kind in London.

Old Jail *Jail Lane, TN16 3AX*
This pub on the very outskirts of London is named after the reported former existence of a building here which served to hold prisoners overnight as they were being transported from London to their final incarceration point of Maidstone. In the pre-train, pre-car era, this journey couldn't be completed in a day.

Old Nick *20-22 Sandland Street, WC1R 4PZ*
This name derives from the fact the current building is apparently on the site of an old police station, with some shackles even found in the cellars.

The Old Nuns Head *15 Nunhead Green, SE15 3QQ*
The story behind this pub also lends its name to Nunhead as a whole. Legend has it that during the Dissolution of the Monasteries in the reign of Henry VIII a nunnery sat on the site of the pub. When the Mother Superior refused to budge, the King's men killed her and put her head on a spike for the community to see. Allegedly her actions saved the rest of the sisters who escaped in an underground tunnel to nearby Peckham.

Old Shades.

Old Tom's Bar.

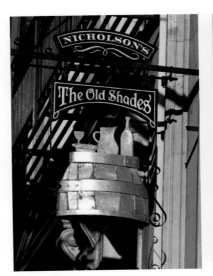

Old Shades *37 Whitehall, SW1A 2BX*
The word shades used to be a generic term for cellars, the pub was originally just called Shades until it acquired its 'Old' prefix back in 1898.

Old Tea Warehouse *6 Creechurch Lane, EC3A 5AY*
So named because it sits on the site of a tea warehouse which once belonged to the East India Company. It was converted to a pub in 1996 where they allegedly found bags of tea under the floorboards which were providing insulation.

Old Tom's Bar *10-12 Leadenhall Market, EC3V 1LR*
Old Tom was a gander who managed to survive being killed for a 19th century banquet and instead ended up having free run of Leadenhall Market, becoming a frequent visitor to the area's many pubs. He ended up living to the ripe old age of 37 and, on his death, even had his own obituary in *The Times*, all of which are ample qualifications for having a pub named after him!

Oliver Conquest *70 Leman Street, E1 8EU*
Takes its name from the man who built the Garrick Theatre on this site in 1830; the pub was included at the back of the building. At the time it was called The Garrick Arms for obvious reasons. Conquest's full name was actually Benjamin Oliver Francis Wyman but you can see why the pub went for his moniker. He also had a career as a moderately successful comic actor and ended up as landlord of the pub for a spell in the 1830s. The theatre was demolished in 1890 but the pub survived. Interestingly this Garrick Theatre predated its more famous namesake, which is also still standing, by nearly 60 years.

One Over the Ait *8 Kew Bridge Road, TW8 0FJ*
This one's a play on the old saying, one over the eight, referring to someone who is drunk. Here it's been modified to Ait which is a name given to islands on the River Thames, as the pub is right opposite Brentford Ait.

One Over the Ait.

The Owl & Hitchhiker.

Open Page *573 Garratt Lane, SW18 4ST*
Another former Front Page pub, by the time this one opened the company's founders were struggling to find any more 'Pages' to name their venues after. Added to that, the landlords of several of their pubs had pushed back on some of their suggestions. For here, they went for this title as they thought it sounded friendly and welcoming. It has certainly stood the test of time as the pub was renamed for a spell after changing hands but then it was decided to turn back to the Open Page.

Orange Tree *45 Kew Road, Richmond, TW9 2NQ*
This takes its cue from the fact the first Orange Tree in the UK was thought to have been planted in Kew Gardens in the late 18th century. The upstairs room of this pub hosted the Orange Tree theatre from 1971, a concept that proved so popular the theatre has now moved a short distance to its own premises.

Overdraft Tavern *200-202 High Street, East Ham, E6 2JA*
London certainly has its fair share of banks turned into pubs, many with fairly nondescript names; however, this former Barclays in East Ham establishment has made a far wittier contribution to the genre. Hopefully it's not a reflection on the impact their drinks prices will have on your bank balance.

The Owl & Hitchhiker *471 Holloway Road, Archway, N7 6LE*
A meeting of two tales in the name of this pub. The owl is from *The Owl & The Pussycat*, Edward Lear's famous poem. A nice touch, as the pub used to be called The Edward Lear. The Hitchhiker in this story comes from Douglas Adams' *The Hitchhiker's Guide to the Galaxy*.

The Owl & The Pussy Cat *34 Redchurch Street, E2 7DP*
This pub takes its name from the 1871 poem by Edward Lear.

Oyster Rooms *Unit 3, First Floor, Fulham Broadway Centre, SW6 1AA*
During the interwar period there was a restaurant that stood on the site of where the pub now stands that was called the Oyster Rooms and Wetherspoons have chosen to pay homage to that in this pub within the small shopping centre that surrounds the entrance to Fulham Broadway tube.

Oyster Shed *1 Angel Lane, EC4R 3AB*
The inspiration here comes from the adjacent passageway called Oystergate walk which in turn is derived from the presence of Fishmongers Hall.

Packhorse and Talbot *145 Chiswick High Road, W4 2D*
This began life as simply the Pack Horse, it had a spell as the 'Upper Pack Horse' before the Talbot was added in 1812. The name is thought to refer to when horses were used to transport goods, known as packhorses, and were accompanied by talbots, a type of dog which would have acted as a guard for the group on their journey. As Chiswick High Road was a key link between London and the west of

Packhorse & Talbot.

England as far back as the 17th century, it is likely it saw plenty of packhorse traffic. However the accuracy of the talbot addition to the name has been questioned in some quarters as they were dogs used for hunting as opposed to guard duty. Either way its inclusion is a helpful addition to avoid confusion with another pub called the Old Pack Horse also on Chiswick High Road. This was why this pub was briefly known as the 'Upper Pack Horse' and the other took the 'Lower' prefix.

The Palm Tree *127 Grove Road, Bow, E3 5BH*
This pub still remains in the middle of Mile End Park, right next to Regent's Canal. An amazing time capsule which hasn't changed in decades, and nor should it, its name comes from the fact Palmer's Wharf used to be situated on the other side of the canal, possibly where palm wood would have been imported to make furniture in nearby Shoreditch.

Partridge *194 High Street, Bromley, BR1 1HE*
A-ha! Another Fullers Ale and Pie House, which means another pub named after a pie.

Pavilion *Wood Lane, W12 0HQ*
Began life as the Rifle Pavilion on account of its proximity to a rifle range on Wormwood Scrubs. By the early 20th century the Rifle prefix was dropped from the pub's name and the facility it was named after was replaced in the 1960s by the West London Stadium, an athletics ground. It survives to this day, now known as the Linford Christie stadium, a name which has endured despite its namesake's fall from grace.

Pavilion End *3 Watling Street, EC4M 9BR*
Cricket themed pub in the heart of Central London; when it was refurbished and taken over by Fullers in 2018, they had former England batter Mike Gatting doing the honours at its official reopening. The owners during the 1980s really played on that theme with the staff wearing cricket whites, a policy which thankfully has been dropped, not least given the garish outfits now worn in limited-overs versions of the game.

Paxton's Head *153 Knightsbridge, SW1X 7PA*
This Grade II listed pub dating from 1851 was named after Joseph Paxton, architect of the Crystal Palace which first stood in Hyde Park when it was home to the Great Exhibition of 1851.

The Peasant *240 St John Street, EC1V 4PH*
If you cast your mind back to history lessons at school, hazy memories of learning of the peasant's revolt of 1381 may emerge. The leader of the uprising, Wat Tyler, met young King Richard II close to the current site of this pub, at Smithfield, to set out his demands, but was attacked and killed by William Walworth, then London Mayor of London.

Pelt Trader *Dowgate Hill, EC4N 6AP*
This pub is opposite a city livery company building which once traded furs, with pelt being an alternative word used for animal fur.

Penderel's Oak *283-288 High Holborn, Holborn, WC1V 7HP*
Named after Richard Penderel, who helped King Charles II escape capture in the aftermath of the Civil War. He helped disguise the King and smuggled him to his brother's estate in Shropshire where he famously hid in an oak tree to escape capture. (See also Royal Oak).

Penny Farthing *3 Waterside, Crayford, DA1 4JJ*
Another one of the rapidly growing collection of micropubs in south-east London, this building was previously used as a bike shop, hence being named after the first vehicle to be called a bicycle, with its incredibly large front wheel!

The Peasant.

The Pensioner *28 Bazely Street, E14 0ES*
Originally called the Greenwich Pensioner on account of the fact navy veterans used to live in Greenwich Hospital, now the Naval College buildings, much in the same way that army veterans lived, and indeed still do live, at Chelsea Hospital (see The Chelsea Pensioner). The name was shortened at the tail end of the last decade.

People's Park Tavern *360 Victoria Park Road, E9 7BT*
In the nineteenth century, Victoria Park picked up a reputation for being the 'People's Park' as for many East End workers it was the only large space of greenery they knew.

Pepper St Ontiod *21 Pepper Street, E14 9RP*
Certainly one of the most peculiarly named pubs in London. However, St Ontiod was not some 4th century monk: the pub is in actual fact an acronym of the address. Ontiod is short for 'On The Isle of Dogs' where the pub can be found on Pepper Street.

The Phoenix *Windsor Walk, Denmark Hill SE5 8BB*
The pub here rose from the ashes of a massive fire in 1981 that gutted the Victorian-era station at Denmark Hill, hence the Phoenix name. When the station restoration was completed in 1986 the old ticket hall building began a new life as a pub, the Phoenix and Firkin, then having a spell as an O'Neills before settling down into its current guise. The fact there was a restored building (Grade II listed, no less) for the pub to move into was considered a sign of a shift of attitudes at British Rail, as their previous stance when station buildings were damaged was usually demolition to be replaced by a flat-pack modular design.

The Pig and Butcher *80 Liverpool Road, N1 0QD*
This harks back to the mid-19th century when the area around the pub was fields used to rest and feed livestock before taking them on their final journey to Smithfield Market.

The Pilot *56 Wellesley Road, W4 4BZ*
While this pub is located under the Heathrow flightpath, this name has nothing to do with aviation. Instead, it's a reference to a crucial navigation tool for a much older form of transport, namely pilot horses for boats sailing down the Thames, which were once stabled here.

The Pineapple *51 Leverton Street, NW5 2NX*
This Kentish Town pub is so named because Leverton Street used to be the site of an exotic fruit market. Ships from India would bring these strange and new fruits to the UK before being taken by traders using wooden carts to the wealthy parts of London to be sold. Having a pineapple would be a sign of your wealth. Legend has it that the garage round the back temporarily housed London Zoo's elephants and giraffes when the zoo's construction was delayed in the Victorian era.

Pipemakers Arms *58 St Johns Road, Uxbridge, UB8 2UR*
A nod to a trade which once operated on this site. Old clay pipes have also been found during digs in the area.

The Pipe Major *1 Yew Tree Avenue, Dagenham, RM10 7FN*
This new pub takes its name from the leader of the Dagenham Girl Pipers which were formed in 1930. Claiming to be the world's first all-female pipe band, they performed throughout Europe before the outbreak of the Second World War. They reformed after the war and are still performing to this day.

Plaquemine Lock *139 Graham Street, N1 8LB*
Appropriately named, given its proximity to Regent's Canal, this New Orleans themed pub is actually named after a real lock on the Mississippi River in the USA. The pub is owned by the great grandson of Jacob Hortenstein (who built the lock) and Carrie B Schwing who opened the lock in 1909. One wonders if the Island Queen steamships (whose London pub namesake is just over the other side of the canal) would have sailed past the locks in days gone by?

Plumbers Arms *14 Lower Belgrave Street, SW1W 0LN*
Owes its title to the fact a local group of plumbers successfully faced down the master builder Thomas Cubitt when he wanted to clear the site as part of his development of Belgravia in the mid-1820s. It acquired a rather grisly claim to fame in November 1974 after a bloodied Lady Lucan stumbled in here after her husband had just killed their nanny and subsequently tried to kill her. Lord Lucan was never found and the mystery of his whereabouts and fate gripped the public consciousness in Britain for years afterwards.

The Pocket Watch *434 Uxbridge Road, W12 0NS*
Originally known as the British Queen, this establishment had a slightly chequered past, so in 2015 owners Brakespear decided to call time on that era and rechristened the pub after a common accessory for Victorian gents. Their thinking behind it was a desire to both reflect tradition while also picking something that would visually lend itself to representation on a pub sign. They certainly managed to find something unique as it's the only pub in the country with this name.

The Pocket Watch.

The Porcupine.

The Prince Blucher,

The Pommelers Rest *196-198 Tower Bridge Road, SE1 2UN*
A pommeler is someone who works with leather. This area just south of Tower Bridge used to be famous for its leather making, so the pub's name is a nod to the heritage of the area.

The Porchester *88 Bishop's Bridge Road, W2 5AA*
The present name dates from 2008 and comes from the nearby Porchester Square and Hall (a wonderful combination of library, venue and Turkish baths!); however, it is its original name which earns its place in history. It was the Royal Oak which gave the nearby tube station its name, putting it in the elite group of six pubs to hold that honour.

Porcupine *48 Charing Cross Road, WC2N 0BS*
This has been a prickly one to interpret: despite being a pub with a rich history, there has been very little written about its name. Two suggestions made in the *Dictionary of Pub Names* from the 1980s were that it could either refer to machinery used to mash beer, which could be considered like a Porcupine, or be taken from any number of coats of arms which feature the creature. The pub was a haunt for the Freemasons during the late 18th and early 19th centuries; perhaps the story behind the name here is another secret they're keeping under wraps...

Porter & Sorter *Station Road, East Croydon, CR0 6BT*
This pub combined the names of two occupations that you would find working nearby. A porter works on the railways helping passengers at stations. Meanwhile a sorter works in a post office sorting the mail. This pub is located next to East Croydon station and there used to be a mail sorting office in the area.

The Porterhouse *21-22 Maiden Lane, WC2E 7NA*
A Porterhouse is an old name for a house where malt liquor, such as porter, is sold. This pub claims to be one of the largest in London as it's on multiple levels. It's located on the site of JMW Turner's house and a plaque inside the pub commemorates this.

The Postal Order *31-33 Westow Street, SE19 3RW*
This pub sits on the site of Norwood Chief District Post and Telegraph office; it was gradually downgraded over time and by the 1950s it was a smaller sorting office before being converted into a Wetherspoons.

Pratts & Payne *103 Streatham High Road, Streatham, SW16 1HJ*
A pub that combines the names of two very different local legends. Pratts was at one time the main department store on Streatham High Street. Cynthia Payne was the owner of a local brothel who was sent to prison after the police raided in 1978 and found 53 men, including a peer, an MP and several vicars. She became a national celebrity in the process as her stories made headlines multiple times during the 1980s. Her business was eventually shut down but she would end up running for Parliament on two occasions, one of which was for the 'Payne & Pleasure' party.

The Pregnant Man *40 Chancery Lane, WC2A 1JA*
Located underneath advertising juggernaut Saatchi & Saatchi's headquarters, the name references one of the company's most successful and memorable ad campaigns from the 1970s, that of a pregnant man with a caption underneath stating: 'would you be more careful if it was you who got pregnant?' for the Family Planning Association. When Saatchi & Saatchi were based in Charlotte Street the Pregnant Man was based within their office block and accessible by their staff only. When they moved to Chancery Lane they kept the idea of a pub for their staff, but this time put it on the ground floor and opened it to all!

Pride of Spitalfields *3 Heneage Street, E1 5LJ*
Known as the Romford Arms up until the mid-80s, the current name is an easy one to interpret: when the landlord started taking beer from Fullers, he took part of the title of their most popular beer and added it to the name of the pub's surrounding area. For many years it also had the honour of being home to one of London's top pub cats, Lenny.

Prince Blucher *124 The Green, Twickenham, TW2 5AG*
This Twickenham hostelry first opened its doors in the early 19th century, shortly after the Battle of Waterloo, and it is in that patriotic fervour that the pub got its name. Ludwig Blucher was head of the Prussian Army and was defeated by Napoleon at the Battle of Ligny. However, he'd made a promise to support Wellington at Waterloo and once he and his army regrouped they returned in time to provide crucial assistance in that final battle.

Prince Frederick *31 Nichol Lane, BR1 4DE*
The only pub in the UK to be named after Prince Frederick, eldest son of George II and so heir to the throne until his death in 1751, nine years before that of his father. It first opened just 10 years after his death as 'Prince Frederick's Head'; the current building dates from the Victorian era.

Prince of Teck *161 Earls Court Road, SW5 9RQ*
This pub dates back to 1868, two years after Prince of Teck, Francis von Hohenstein, married into the British Royal Family via Princess Mary Adelaide of Cambridge, a granddaughter of George III. It also sounds a bit like one of those slightly dubious electronic shops that used to be at the Centre Point end of Tottenham Court Road.

Princess Louise *208 High Holborn, WC1V 7EP*
Noted for its Grade 2 listed urinals and its restored interior which brought back old-fashioned dividers, this pub is named after the sixth child of Queen Victoria and Prince Albert. Born in 1848, she became a successful sculptor and many of her works are still visible today. When her husband was appointed Governor General of Canada she began an association with the country, with the province of Alberta, Lake Louise and Mount Alberta also named after her, along with the pub.

Princess of Prussia *15 Prescot Street, E1 8AZ*
Named after Princess Charlotte of Prussia, born in 1860, the same year the pub opened. Like the earlier entry of Prince Blucher in Twickenham, another demonstration of the close links between Britain and the Prussians in the 19th century, a situation that reversed dramatically with the onset of the First World War. Interestingly there is no reference to this pub changing its name in that period, which was the case with other establishments with Germanic links.

Prospect of Whitby *57 Wapping Wall, Wapping, E1W 3SH*
Another of Wapping's riverside pubs and arguably one of the most famous in London, it was originally named Devil's Tavern after picking up a notorious reputation by the 17th century. It has been a tavern since 1520 but after a refurb around 200 years ago the landlord decided to rename it the Prospect of Whitby after a collier ship that regularly moored outside the pub which carried coal between Newcastle and London.

Punch & Judy *40 Henrietta Street, WC2E 8RF*
This Covent Garden pub is named after the famous puppet show which was performed nearby. Indeed, Samuel Pepys wrote about the first performance in his diaries in 1662.

The Punch Tavern *99 Fleet Street, EC4Y 1DE*
Renamed in the 1840s after the famous satirical magazine which was based nearby. The magazine is no longer with us but the pub lives on.

Princess of Prussia.

Punch & Judy.

The Pyrotechnists Arms *39 Nunhead Green, Nunhead, SE15 3QF*
Many pubs with 'Arms' often refer to the occupation or hometown of
the landlord, which is why many pubs with this sort of name didn't
make the cut for this book. However, this pub in Nunhead is on the site
of a former fireworks factory, hence its name.

The Queens Head and Artichoke *30-32 Albany Street, NW1 4EA*
I, (Sam that is!) have a confession to make with this pub. When I first
moved to London in 2006 and visited the pub, I thought the name was
one of those examples of where a landlord had tried to jazz up an old
pub name by whacking in a random item at the end. It has, however,
been called this for well over 150 years. There are two theories
circulated as to why, both involving Tudor monarchs. One contends
that the first landlord of a pub on the site was Daniel Clarke, head cook
to Queen Elizabeth and King James, and also their chief gardener so
would have both grown and cooked that vegetable. The other is that
Henry VIII's sister was a big fan of artichokes and managed to persuade
the landlord at the time to add them to the pub's name. Whichever of
the two it is, what is certain is that it's a good example of not judging
a book by its cover!

The Queens Larder *1 Queen Square, Holborn, WC1N 3AR*
The madness of King George III explains the name of this pub. When
news of his illness was still a secret, he was staying at his physician's
house on Queen Square. It would be his queen Charlotte who hired a
cupboard in the nearby pub to store provisions to help him through his
isolation. When this became public knowledge, the pub was named in
honour of this story.

Quinns *65 Kentish Town Rd, NW1 8NY*
This Camden pub, with its distinctive yellow walls, has had this name since 1991, when it was acquired by the Quinn family who still run things today.

Rabbit Hole *151-153 Greyhound Lane, SW16 5NJ*
Now this is the *Alice in Wonderland* themed pub we promised you earlier! Previously called the Greyhound after the road it's on, the place is now decorated both inside and out with a variety of motifs from the *Alice in Wonderland* universe, with a particular emphasis on our long-haired furry friends. Size wise, it's a cavernous pub both inside and out so the name is apt in that regard too.

The Queens Larder.

Racketeer *105 King's Cross Road, WC1X 9LR*
A relatively recent renaming of the former Carpenter Arms, the tiled signage with the original name remains outside. The name racketeer was chosen as a nod to the Dickensian past of the local area, with racketeers in that era being individuals who would create a distraction which would then allow the criminals to swoop in and pickpocket their victims.

Radicals & Victuallers *59 Upper Street, N1 0NY*
This pub near Angel provides us with a joining of two pieces of local history. The Radicals refers to the Labour Party, who have had a strong presence in Islington and which was also at the heart of the punk scene in the 1980s. A Victualler is someone who is licensed to sell alcohol and the connection here is that Islington, certainly at one time, contained the highest number of licensed premises in the country.

Rack & Tenter *45 Moorfields, Tenter House, Moorgate, EC2Y 9AE*
This pub's name comes from the fact Moorfields used to be a hub for the textile industry. Racks and tenters were part of the equipment the workers would use to make the fabrics.

The Railway Telegraph *112 Stanstead Road, SE23 1BS*
Opened in 1853, The Railway Telegraph is so named from the fact the London, Brighton & South Coast Railway installed a telegraph system along its route to London Bridge in the same year.

Randall Tavern *Field Way, New Addington, Croydon, CR0 9DX*
This was originally known as The Bunkers Knob which, before anyone gets any ideas, was named after William Coppin, a policeman who had a clubfoot which made a bonking noise when he walked, which gave him the nickname Bunker. He spent a lot of time on a nearby hill which in turn was called Bunkers Knob. It was renamed in 1998 in memory of Derek Randall, a landlord here for eight years, who sadly passed away in March of that year.

The Red Lion *Various - Twenty Three Pubs in London*
One of the country's most popular pub names, with many found in the capital, the Red Lion can trace its name to John of Gaunt who was a key figure in 14th-century England. He was son of Edward III and nephew to Richard II and his own son would become Henry IV. He left England after the crown passed to his nephew to take up the throne of Castile, where his wife came from, and incorporated the red lion from Castile's coat of arms into his own. This led to many taverns showing a red lion in support of John as Richard was unpopular after ending the Peasants Revolt, led by Jack Straw. Richard responded by demanding that pubs display his coat of arms of a White Hart (see also White Hart) instead. However, John would return to England to support his nephew and stabilise the country.

Red Barn *Barnehurst Road, DA7 6HG*
This is thought to be a reference to the fact the area where this pub and the housing surrounding it now stand was once part of Longlanes Farm, which sprawled over a considerable area until this south-east London suburb developed in the early 20th century. The Red Barn became a key venue for British jazz in the immediate post-war period and included performers such as Humphrey Lyttelton who, as we know, now has a pub named after him in Mornington Crescent.

Red Lion and Pineapple *281 High Street, Acton, W3 9BP*
Here we have two pubs that were fused together as part of a road widening scheme in 1906. They may no longer exist as separate entities but their legacy has been preserved by this name.

Reliance *336 Old Street, EC1V 9DR*
Previously a chicken feed shop with a massive model chicken looming over Old Street, much of the pub's interior was constructed from an old Thames sail barge called Reliance.

Resting Hare *8 Woburn Walk, 11 Upper Woburn Place, WC1H 0JW*
This charmingly evocative name for this Bloomsbury pub seems rather out of character with the heavily built-up area it inhabits. However, when Woburn Walk was first built, a resting hare was a common sight on the pavement here at night and the pub's own website cites the renowned poet W.B Yeats, a former resident of the walk, once writing

of seeing a 'handsome old grey hare' resting outside No.6 here in the 1920s. Sadly, the hares have long gone, as probably has the prospect of anyone trying to do anything resembling resting in the vicinity of the Euston Road juggernaut.

Richmal Crompton *Westmoreland Place, BR1 1DS*
Named after the author of the *Just William* books, Crompton penned the first short stories featuring William Brown while working as a teacher at Bromley School for Girls.

The Resting Hare.

The Rifleman *7 Fourth Cross Road, TW2 5EL*
This pays homage to the fact that riflemen were stationed nearby in the Napoleonic era.

The Ring *72 Blackfriars Road, SE1 8HA*
This pub takes its name from a curiously shaped former church that was converted into a boxing arena in 1910 and which would be right opposite the pub if it still stood today. When first built as the Surrey Chapel in 1783, the preacher behind it, Reverend Roland Hill, had it designed in such a way there were no corners in the building for the devil to lurk. It was converted into a boxing venue by Dick Burge, a former British lightweight champion, and did a roaring trade until being hit twice during the blitz and demolished. The site is now occupied by the Palestra office block.

Rising Sun *Various - Eighteen Pubs in London*
Several pubs in London share this name, mainly for its association with hope and optimism. However, it also features on the crest of the Worshipful Company of Distillers, one of the City of London's livery companies, making it a popular pub name.

Robin Hood and Little John *78 Lion Road, DA6 8PF*
On first glance this looks like a homage to Robin Hood by referencing its main character and one of his merry men. However, the Little John in question is said to be the son of the licensee of the pub during the second part of the 19th century, when it was simply called Robin Hood, who was killed when a dray (a truck without sides which delivers beer barrels) accidentally reversed back onto the poor young lad.

The Rochester Castle *143-145 Stoke Newington High St, N16 0NY*
Built in 1801 by Richard Payne, who was from Rochester.

The Rocket *Various - Four Pubs in London*
A few pubs in London commemorate Robert Stephenson's locomotive which ran the world's first passenger train between Manchester and Liverpool in 1829. It was preserved in the Science Museum until 2018, when it was moved to the National Railway Museum in York.

The Rockingham Arms *119 Newington Causeway, Elephant and Castle, SE1 6BN*
Charles Watson-Wentworth was the second Marquess of Rockingham who was briefly Prime Minister on two occasions, once in the 1760s and again in the 1780s. The Wetherspoons in Elephant & Castle is named in his honour.

Rookwood Village *314 Cann Hall Road, E11 3NW*
Previously known as Lord Rookwood, the Lord in question being Selwyn Ibbetson, the MP who was elevated to the House of Lords as he successfully got the Epping Forest Act through the Commons in 1878. The act meant the area was protected from developers and became an open space for the population to use freely. The village aspect comes from the pub company that now owns it.

The Rose & Crown *Various - Eighteen Pubs in London*
At the conclusion of the Wars of the Roses in the 1480s, the new King Henry VII decided to finally end the possibility of a continuation of the wars by marrying Elizabeth of York. With both houses united, Henry merged both red and white roses to give the Tudor Rose. In wishing to display loyalty to the new King, a number of innkeepers and landlords decided to display the new rose with a crown.

The Rosemary Branch *2 Shepperton Road, N1 3DT*
This pub can trace its name to the Levellers movement during the English Civil War. Levellers would use a branch of rosemary or a twig to designate their affiliation. There have been at least three pubs with this name on this site near Regent's Canal, with the current building dating from the late 19th century.

Round Table *26 St Martins Court, WC2N 4AL*
This dates back to the 18th century when several of the pubs on St Martins Lane were known as places for fighting men to meet and arrange bouts with each other. The round table was adopted here as the pub sign to indicate that even handed justice would be observed within the premises. The current pub building dates from 1877 so was not the one used during the brawling heyday.

The Royal Oak *Various - Nineteen Pubs in London*
One of the most popular pub names in the country, the tree in question was located at Boscobel House in Shropshire where King Charles II hid whilst on the run from Cromwell's forces in 1651. The reason why this

name is so popular is that it was used to demonstrate support for the
monarchy when Charles was restored to the throne in 1660. Royal Oak
is also a station on the London Underground, although the pub there
has since changed its name (see The Porchester).

The Round Table.

Salmon & Ball.

Royal Standard *Various - Nine Pubs in London*
The Royal Standard is the flag used by the monarch whenever
they are present in a building. The current flag is divided into four
quadrants, the first and fourth are three gold lions on a red background
representing England, the second a single red lion on a gold background
for Scotland and the third a gold harp on a blue background for Ireland.
One wonders if these pubs have a flag they would raise should the
monarch ever visit!

Rugby Tavern *19 Great James Street, WC1N 3ES*
This one has nothing to do with sport, the name derives from the fact
the pub and the surrounding area of housing was all built on land
originally owned by Lawrence Sheriff who set out that, upon his death,
it should be used as an endowment to realise his vision of creating a
school for the local area where he grew up in Rugby. The school still
owns 103 properties in this part of London so it remains a lucrative
source of income.

The Running Horse *Various - Three Pubs in London*
There are a number of stories vying for contention on this one. The pub
in Erith is named after the wild horses that would have roamed on the
marshes nearby, but other pubs in the country hint to a more general
connection with horseracing, making this story more of a photo finish.

Rusty Bucket *11 Courtyard, SE9 5PR*
A distinctly cheeky name for this modern micropub that opened in 2018. The owners originally wanted to go all out with the cock and bucket; however, on reflection, they decided to tone it down a little and plumped for the Rusty Bucket. It's their own little bit of cockney rhyming slang: we'll leave you to figure it out!

Salmon and Ball *502 Bethnal Green Road, E2 0EA*
A few theories abound regarding the story behind this distinctive pub located slap-bang opposite Bethnal Green tube. A common one is that it references two trades which were in their heyday when the pub opened in 1733, namely 'Salmon' for Billingsgate fish markets and 'Ball' for a ball of yarn for Weavers Field, which is where cottages housing people working in the silk weaving industry once stood. Others have suggested the ball was chosen because it was the emblem of Constantine the Great and the silk weavers adopted it themselves because silk came from the east. The pub was renamed as Tipples in the 1980s but, thankfully, in 1991 the original name was reinstated.

The Salutation *154 King Street, Hammersmith, W6 0QU*
Named after the religious greeting associated with the annunciation. In living memory the pub's sign has shown two different types of salutation, at one stage showing a soldier saluting a military cross. Its present design shows one of the courtiers of Elizabeth I bent on one knee, offering salutations of a different nature to his Monarch.

Samuel Pepys *Stew Lane, EC4V 3PT*
Samuel Pepys was the famous diarist who provided us with a huge insight into life in 1660s London. He would later become an MP and his account of his times is an invaluable resource for historians. He also famously buried his cheese during the Great Fire of London in an attempt to stop it from melting.

Sanctuary House Hotel *33 Tothill Street, SW1H 9LA*
Hundreds of years before the building which now houses this pub and hotel was built, this site was used by monks from Westminster Abbey to provide food, drink and shelter to the poor or, in other words, offer sanctuary, hence the name today.

The Saxon Horn *352 Upminster Road, RM13 9RY*
Our second archaeologically inspired name, this marks the fact a Saxon drinking vessel was found in Rainham; it was previously on display in the British Museum but is currently in their store.

The Saxon King *198 Petersfield Avenue, RM3 9PP*
The entire area of Harold Wood is named after the defeated Monarch at the Battle of Hastings; this pub name continues that theme by recognising his status as the final Saxon King of England.

The Scarsdale Tavern *23a Edwardes Street, W8 6HE*
Named after a former grand old house of the same name that was based nearby, close to where Kensington High Street is today, and was owned by the Curzon family, whose hereditary title was Baron Scarsdale. The pub is also a few streets away from Scarsdale Villas which date from the mid-Victorian period, and are also thought to be named after the house and built around the same time as the pub.

The Scolt Head *107a Culford Road, N1 4HT*
When sister and brother team Rosie and Rich took over what was the Sussex Arms in the mid-noughties they didn't have the money to paint the exterior of the building, but still wanted to make it clear to the community the pub was under new management and worth a look. As a result, they decided to change the name, initially thinking of keeping it as a county and mulling over Norfolk as it was where they'd spent their family holidays each year. However, they were mindful of the potential for strong Alan Partridge connections and decided to make it a bit more specific, choosing Scolt Head, an island off the north Norfolk coast.

The Scottish Stores *6 Caledonian Road, N1 9DT*
The pub first opened in 1901 when the building was also used to store meat transported from Scotland on the East Coast Main Line down to King's Cross. It gained infamy during its time as the Flying Scotsman when the venue became a 'strip pub', complete with blacked out windows and toilets which one online reviewer commented were so bleak they reminded him of the ones featured in the film 'Trainspotting'.

The Seven Stars *53 Carey Street, Holborn, WC2A 2JB*
A survivor of the Great Fire of London, this pub dates from 1602. Its name comes from the fact that Dutch sailors allegedly used to drink there and it was originally named the League of Seven Stars. This is to represent the seven provinces that make up the Netherlands.

The Shaston Arms *4-6 Ganton Street, W1F 7QN*
A nod to the home county of the brewery that owns the pub, Hall and Woodhouse, who are based in Blandford St Mary, Dorset. Shawston was the name used for the Dorset town of Shaftesbury by arguably the county's most famous author, Thomas Hardy. Picking it as the name gave the perfect opportunity to signify this Soho pub is a little bit of the country in the heart of the city!

Shaw's Booksellers *31-34 St Andrew's Hill, EC4V 5DE*
Before becoming a pub in 1997 this building housed a paper mill and was particularly well located to serve what was once the heart of Britain's newspaper industry, Fleet Street. The building was mocked up as a book shop for the 1997 film *Wings of a Dove*, called Shaw's Booksellers, and the name has stuck!

Shaws Booksellers.

Sheephaven Bay *2 Mornington Street, NW1 7QD*
Named after a charming spot in the landlord's home county of Donegal on Ulster's north coast.

Shelverdine Goathouse *7-8 High Street, SE25 6EP*
Popular with Crystal Palace fans as their ground is nearby, the Goat House used to be a landmark pub in the area, and South Norwood (where the pub can be found) used to go by the name of Shelverdine.

Sherlock Holmes *10 Northumberland Street, WC2N 5DB*
Originally called the Northumberland Arms, the pub was renamed after Sir Arthur Conan Doyle's world-famous detective in the 1950s as the pub had featured in his 1892 short story, *The Adventure of The Noble Batchelor*, and subsequently acquired a treasure trove of Holmes memorabilia

Shinner & Sudtone *67 High Street, Sutton, SM1 1DT*
Another one combining two local terms, this pub in Sutton is named after Shinner's Department Store which opened in 1935 and Sudtone is an old name for Sutton, first mentioned in 675 AD.

Ship Aground *Wolseley Street, SE1 2BP*
This pub is located by a part of the Thames where many ships ended up running aground; it also featured in the late 80s to mid-90s ITV drama *London's Burning*, which may well be showing on ITV3 or London Live while you're reading this entry!

The Ship & Shovell *1-3, Craven Passage, Charing Cross, WC2N 5PH*
Originally named the Ship & Shovel (with one L) as Charing Cross
used to be a dock and workers would have to shovel off the goods
arriving in London. But it was renamed by adding an extra L after a
famous admiral, Sir Cloudesley Shovell. He served during the Nine
Years' War and the War of the Spanish Succession. However, his career
was to end in disaster when his ship and three others ran aground
on the Isle of Scilly. An estimated 1,400-2,000 sailors lost their lives,
including Shovell, making it one of the worst disasters in the history
of the Royal Navy.

The Ship & Shovell.

The Ship & Whale.

Ship and Whale *2 Gulliver Street, SE16 7LT*
Like the Moby Dick, takes its name from the whaling industry which
was one of the early users of Greenland Dock during most of the 18th
century.

Shy Horse *423 Leatherhead Road, Chessington, KT9 2NQ*
Originally called the Fox and Hounds, the pub was renamed The
Shy Horse in 2003 in support of the ban on fox hunting, with a Shy
Horse being known in the hunting world as a horse that refused to get
involved in the chase.

Silver Fox *9 Montpelier Avenue, DA5 3AP*
Another witty contribution from our cohort of south-east London
micropubs, this one stems from the fact one of the co-owners has silver
hair and the other's surname is Fox.

Simpsons Tavern.

Simpsons Tavern *38½ Cornhill, EC3V 9DR*
Named after Thomas Simpson, who first got his name about town in the 18th century when he opened a successful fish restaurant in Bell Alley near the market at Billingsgate. In 1757 he opened his chop house here after the land was gifted to him by his father. It's been going strong ever since, proving having a half in your address is no barrier to success!

Simon the Tanner *231 Long Lane, Bermondsey, SE1 4PR*
As we discovered earlier with the Leather Exchange, the Bermondsey area was once a hotbed of the tanning industry (the name for the process that takes the animal hides to create leather, as opposed to the type popular with reality TV stars). Simon the Tanner is thought to refer to the patron saint of this trade from the Coptic Orthodox Church, who was a shoemaker and is said to have moved Mokattam Mountain in Cairo in the 10th century. There is now a monastery and church named after Saint Simon carved into Mokattam Mountain, which featured in Levison Wood's Channel 4 Documentary *Walking The Nile*.

Sindercombe Social *2 Goldhawk Road, W12 8QD*
Named after Miles Sindercombe, a leveller who plotted to assassinate Oliver Cromwell as he felt he had strayed from the ideals of a puritanical republic. Sindercombe had a cottage in Shepherds Bush near where the pub stands. He attempted multiple unsuccessful attempts on the Lord Protector's life until he was finally caught in January 1657. He was found guilty of treason a month later but took poison to avoid the humiliation of being hung, drawn and quartered. As a puritan, we imagine he wouldn't be best pleased by a pub being named after him.

Singer Tavern *1 City Road, EC1Y 1AG*
This pub is based on the ground floor of an old office building once used by the Singer Corporation, manufacturer of sewing machines. The pub has incorporated one of their old models, much like the ones the clothes shop All Saints used in their shopfronts, into their logo.

Sir Alfred Hitchcock Hotel *145 Whipps Cross Road, E11 1NP*
Taking its name from one of Leytonstone's most famous residents, the pub and hotel here first opened in 1981, a year after the great director's death. Hitchcock is also commemorated by a series of mosaics depicting several of his films in the corridors of Leytonstone tube station.

Sir Christopher Hatton *4 Leather Lane, EC1N 7RA*
Named after the man who built Hatton House in the late Elizabethan era. Hatton Garden was developed nearly 100 years later by his descendants and then developed as the heart of London's diamond trade from the 19th century onwards.

Sir Colin Campbell *264-266 Kilburn High Road, NW6 2BY*
This pub in Kilburn is named after the Scottish army officer in the 19th century. He commanded the Highland Brigade during the Crimean War and would later serve as Commander-in-Chief in India. There is also a pub in Southwark named after his hereditary title, Lord Clyde.

The Sir John Hawkshaw *Cannon Street Station, Cannon Street, EC4N 6AP*
Sir John Hawkshaw was the designer of Cannon Street Station where this Wetherspoons can be found.

The Sir John Oldcastle *29-35 Farringdon Road, EC1M 3JF*
Sir John Oldcastle was allegedly the inspiration for Falstaff who appeared in numerous Shakespeare plays. There was a pub named after him which sat in the grounds of his old mansion, which was nearby.

The Sir Julian Huxley *152-154 Addington Road, Selsdon, South Croydon, CR2 8LB*
Sir Julian Huxley was the first Director-General of UNESCO who advocated for the Selsdon Wood Nature Reserve which was opened in 1936.

The Sir Michael Balcon *46-47 The Mall, W5 3TJ*
Ealing's Wetherspoons remembers the famous film producer who took over the nearby studios in 1938. His Ealing comedies such as *Passport to Pimlico* and *The Ladykillers* became immensely popular after the Second World War. Sir Michael also helped launch the career of a young Alfred Hitchcock in the 1920s.

Sir Richard Steele *97 Haverstock Hill, NW3 4RL*
Built on the site of a cottage which was once occupied by Sir Richard Steele, a writer and playwright as well as an MP when elected to parliament for Stockbridge for the Whigs in 1713. He is most famous for being the co-founder of The Spectator magazine. In 2009 the Steele legacy expanded beyond the pub when local residents rechristened the area between Belsize Park and Chalk Farm stations as 'Steele's Village' as a way to promote civic engagement and boost local businesses.

Sir Sydney Smith *22 Dock Street, E1 8JP*
This holds the rare distinction of being named after a famous person while they were still alive. When it opened in 1809, Sir Sydney, a British Admiral from the Napoleonic Wars, was still going strong and almost as famous as Nelson. As time has passed this colourful character, who was also a strong campaigner against slavery, has faded into obscurity. It is the pub's position close to the then thriving docklands which is thought to have been behind naming it after someone else connected with the high seas.

Skehans *1 Kitto Road, SE14 5TW*
Skehans is one of the few family-run pubs in south London. Bill Skehan took over the pub in the early 2000s.

Sir Richard Steele.

The Skylark *34-36 South End, Croydon CR0 1DP*
Named after Gerard Manley Hopkins' famous poem, 'The Sea and The
Skylark'. (See also The Goldengrove)

The Slaughtered Lamb *34-35 Great Sutton Street, EC1V 0DX*
Another pub based near Smithfields with Lamb in the title, this one
is named after the Yorkshire pub featured at the start of the cult US
1981 horror-comedy, *An American Werewolf in London*. The American
walkers failed to heed Brian Glover's warning to stay away from the
moors, with an inevitably bloody outcome. The closest to a moor
round here is Moorgate which is typically a werewolf-free zone.

Snooty Fox *75 Grosvenor Avenue, N5 2NN*
Originally named the Grosvenor Arms, this was renamed in 2003 after
a club from the 1960s that the co-owner, Stuart, had always wanted to
go to. In 2008 the reigns were handed over to his daughter Nicole who
is still in post today and is keeping the 1960s theme alive with a vinyl
jukebox and pictures of that decade's icons decorating the pub's walls.

The Sparrowhawk *2 Westow Hill, Crystal Palace, SE19 1RX*
A Crystal Palace connection here: when it was first built in Hyde
Park in 1851 it surrounded many historic elm trees. This resulted in
the inadvertent and unwelcome inclusion in the Great Exhibition of
sparrows which had been nesting in those trees before the palace was
built. Queen Victoria was definitely not amused by this; shooting the   The Slaughtered Lamb.

The Spanish Galleon.

birds was out of the question given it was a glass palace. The story goes that it was the Duke of Wellington who came up with the solution: when the Queen asked him what could be done he simply replied 'Sparrowhawks, Ma'am'!

Sylvan Post *Heron House, 24-28 Dartmouth Road, SE23 3XZ*
This pub in Forest Hill is located in the old post office, where some of the original features can still be seen, Sylvan being the Latin for forest.

The Sovereign of the Seas *109-111 Queensway, Petts Wood, Orpington, BR5 1DG*
Built in 1637, the Sovereign of the Seas was King Charles I's flagship. At his insistence the ship had 102 bronze cannons, making it the most powerful ship in the world at the time. It served in numerous battles until 1695 when a fire burnt the ship to its waterline.

Spaniards Inn *Spaniards Road, Hampstead, NW3 7JJ*
No one is quite sure why the Spaniard's has its name. A number of theories suggested include it being run by two Spanish brothers; it was previously the home of the Spanish Ambassador or even the Spanish Embassy itself. One thing is for certain, that this part of north London has had some connection to Spain at some point!

Spanish Galleon *48 Greenwich Church Street, SE10 9BL*
Takes its name from the numerous paintings displayed in the nearby Greenwich Naval College which depict British forces battling against the Spanish Armada.

The Speaker *46 Great Peter Street, SW1P 2HA*
Formerly an Elephant and Castle until 1999, since then it has been The Speaker, reflecting its proximity to the Houses of Parliament. The pub even used to have a division bell (which indicates a vote in Parliament) on the premises. Despite that, attempting to purchase a drink by shouting 'Order, Order!' is not advised...

Spice of Life *6 Moor Street, W1D 5NA*
There are three ingredients to why this Soho pub was renamed in 1986 in honour of a line from William Cowper's 1785 poem, 'The Task'. Firstly, Soho as an area was thought to perfectly embody the idea of 'variety is the spice of life', given everything that is packed into its streets and square. Secondly, it was a nod to one of the communities that have settled in Soho, the Chinese, and their use of spice in their cooking. The final link was with the poet himself, who hailed from Hertfordshire which is also where the McMullens, the brewery that bought and still owns the pub to this day, are based.

Spit and Sawdust *21 Bartholomew Street, SE1 4AL*
The origins of this one start in Ipswich in 2009, where one of the owners visited a new pub serving beer from independent breweries. The pub was a stripped back experience with beer barrels on display; while he was telling his Dad about it, he replied that it sounded like the 'Spit and Sawdust' pubs he used to visit in the 1960s. The phrase stuck in his mind and when, several years later, he and a friend decided to open a pub it was the obvious choice for their vision of a traditional wood finish pub in a contemporary, modern setting.

Spring Grove *13 Bloomfield Road, KT1 2SF*
Takes its name from the former Spring Grove estate, formerly 11 acres of farm and meadowland before it was divided up into various plots for development in 1866; the pub soon followed a year later and today many still refer to this area of Kingston as Spring Grove.

Sporting Page *6 Camera Place, SW10 0BH*
This was the second venue to open in the Front Page chain, a short distance from their first venue with that name. Sporting Page was chosen because the owners felt that's the next page people instinctively turn to, after reading the front page. The Front Page is no more (now the Chelsea Pig); it was named after a French brasserie that the owners both loved and helped pave the way for several more pages to pop across London. Two still remain which we've documented in our earlier pages, but the book has been closed on other members of the family like the Racing Page in Richmond and the City Page by Cannon Street.

The Spotted Dog *15a Longbridge Road, Barking, IG11 8TN*
Legend has it that this pub is sited on one of Henry VIII's old hunting lodges. It then became the home of the King's Master of the Hounds, which would explain its canine connection.

Spouters Corner *180 High Road, Wood Green, N22 6EJ*
This pub is built on a piece of land which was once known as Spouters Corner and used for open-air meetings. It was first used for these purposes for a meeting in support of the Reform League campaign to expand the right to vote to all men, in 1867, and was still in active service up until the 1950s, hosting early CND demos. There are probably still people keeping the tradition of spouting alive here to this day, although perhaps not over such weighty matters.

The Spurstowe Arms *68 Greenwood Road, E8 1AB*
William Spurstowe (1605-1666) was the local vicar in Hackney in the 1640s. He is remembered locally for beginning the construction of six Almshouses in the area before his death.

Square Tavern.

The Star of the East.

Square Tavern *26 Tolmer's Square, NW1 2PE*
Based in the early 1980s redevelopment of Tolmers Square, the original plans to redevelop the Victorian houses in the area sparked massive protests called the 'Battle for Tolmers Square'. The original buildings were still demolished but Camden Council bought the site and the plans were much more modest than the original developer's aspirations of half a million square feet of commercial space.

Stag and Lantern *11-12 The Broadway, Highams Park, E4 9LQ*
The Stag represents local forests in the area and the Lantern was chosen to be symbolic as a beacon of light, representing east London and human industry. This micropub's website adds further context by saying it was chosen to be a throwback to traditional pub names while also being representative of the character of the local area.

The Star of the East *805A Commercial Road, Limehouse, E14 7HG*
This pub which opened in the early 19th century is thought to be named after the Star of Bethlehem, which signalled the birth of Jesus and was spotted by the wise men who were then guided by it from the East to Bethlehem. The name also links back to a period of Limehouse's history, albeit a connection which is coincidental rather than deliberate. From the mid-19th century onwards, London's first Chinatown began to emerge in the area as a result of Chinese sailors working at the docks. The community shifted to Soho by the 1960s as slum clearances and war damage saw the old streets redeveloped by the local council as social housing blocks.

Starting Gate *Station Road, Wood Green, N22 7SS*
As well as being the birthplace of television, the grounds of Alexandra Palace played host to a racecourse which operated for 102 years between June 1868 and September 1970, hence the Starting Gate!

St Brides Tavern *1 Bridewell Place, EC4V 6AP*
This pub takes its name from the nearby church which was rebuilt by Christopher Wren following the Great Fire. However, St Bride was actually founded as St Brigid's who is the patron saint. The church is also associated with journalists as Fleet Street is a stone's throw away.

St Brides Tavern.

The Stonemasons Arms.

St Stephen's Tavern *10 Bridge Street, Westminster, SW1A 2JR*
One of the pubs right beside the Houses of Parliament, this name takes its cue from both St Stephen's Hall and St Stephen's Tower (not the former name of the Elizabeth Tower which hosts Big Ben, despite common assumptions) within the Parliamentary estate.

The Steam Packet *85 Strand-On-The-Green, W4 3PU*
Named after the 19th century steam ships which were moored in the area and used to transport passengers up the Thames, this has only recently returned to being a pub after a long stint as a restaurant.

The Stonemasons Arms *54 Cambridge Grove, Hammersmith, W6 0LA*
This pub was previously called the Cambridge Arms but was renamed in 1997, although working with stone is one of humanity's oldest trades.

Stormbird *25 Camberwell Church Street, SE5 8TR*
An air of mystery surrounds the name of this Camberwell pub as the owners have never provided an explanation behind it, which is said to be a secret held between its manager and her father. One potential explanation shared with us is that it could refer to a racehorse, called Storm Bird, who had a brief but successful career at the start of the

The Sugar Loaf.

1980s before being plagued by illness and injury. There is a picture of Storm Bird the racehorse behind the bar at the Hermit's Cave, which shares the same owners as the Stormbird. Could that be it? I'm not a betting man so we won't lay money on it, but it's a nice story all the same!

The Sugar Loaf 65 Cannon Street, EC4N 5AA
Back in the days when sugar was a more expensive commodity, it would come in the form of a sugarloaf which was conically shaped (imagine something similar looking to a traffic cone). This lasted until the early 1800s when they were replaced by sugar cubes and granulated sugar that we are more familiar with today.

The Sultan 78 Norman Road, SW19 1BT
This post war pub replaced another of the same name which was damaged beyond repair during World War Two. The original opened in 1868 and was named after a successful racehorse from the 1820s.

The Sun and 13 Cantons 21 Pulteney Street, Soho, W1F 9NG
This has its origins in the pub's association with the Swiss watch-making community that worked in the area in the late 19th century. After the original Sun pub was rebuilt in 1882 after a fire, the 'and 13 Cantons' was added to its name as a tribute to the Swiss workers in the area. There were 12 cantons (the Swiss version of county) in Switzerland at the time, so this expat community in Soho got affectionately labelled as the '13th canton'.

The Sun and 13 Cantons.

The Sun in Splendour *7 Portobello Road, W11 3DA*
Takes its name from Edward IVs personal symbol, which is a sun with a human face with a series of waves emanating from it. This was adopted by the King when he was said to witness the fantastical sight of three suns combining together in the sky ahead of a key battle at Mortimer's Cross during the War of the Roses in February 1461. The King took this as an omen of victory which certainly turned out to be the case at Mortimer's Cross, where the Lancastrians were routed.

The Surprise *6 Christchurch Terrace, SW3 4AJ*
The Surprise in question here is HMS Surprise, a French boat originally called Unite which the British captured off the coast of Algeria in 1794 and subsequently renamed. There are several other pubs in London called Surprise, but this is the only one that has nailed its colours to the mast in terms of where its name comes from, so that's why it's made the cut.

The Swimmer at the Grafton Arms.

Sutton Arms *6 Carthusian Street, EC1M 6EB and 16 Great Sutton Street, EC1V 0DH*
In 1611 Thomas Sutton bought the London Charterhouse, at the time an impressive mansion building, with visions of creating a hospital and school on the site. Sadly, Sutton didn't see out the year but his vision for both buildings was realised and so it seems only fitting that not only one but two pubs are named after him in the surroundings of the Charterhouse, part of which still today remains as charitable housing for older people.

The Swimmer at the Grafton Arms *13 Eburne Rd, Holloway, N7 6AR*
The swimmer in question is none other than the owner of Remarkable Pubs, the chain which runs this particular public house. Robert Thomas won the 100 yards national backstroke championship during his school days and decided to name one of his pubs in honour of this achievement. The Grafton Arms part comes from the Duke of Grafton. Of the twelve Dukes to date, the most famous one is the third who was Prime Minister during the 1760s.

Sydney Arms *Old Perry Street, BR7 6PL*
Named after one of the local lords of the manor, John Robert Townsend, the third Viscount Sydney. One of his most significant acts locally was attempting to block the creation of protections in law for Chislehurst common, restricting his powers to exploit the land. He failed in his attempts and the common remains as an open space that can be enjoyed by all.

Sympathetic Ear *37 Tulse Hill, SW2 2TJ*
This represents the atmosphere its owners want this micropub to have, a craft beer venue that is welcoming and inclusive, harking back to community pubs of old (much like the Dodo in Hanwell) and precisely the kind of place you'd find a sympathetic ear if you wanted to share your troubles.

The Tailor's Chalk *47-49 High Street, Sidcup, DA14 6ED*
This Wetherspoons sits on the site of two former businesses, one a dry cleaners, the other a tailor's, which were there for pretty much all of the first half of the 20th Century. Tailors often use chalk to mark where an item of clothing needs altering or adjusting as it can easily be brushed out. Don't be fooled, however, the chalk is actually made of talc!

Tally Ho *749 High Road, North Finchley, N12 0BP*
"Tally Ho" is a term shouted when fox hunting when a fox has been spotted.

Tamesis Dock *Albert Embankment, SE1 7TY*
This pub on a boat takes its name from the old Roman word for the Thames. It's a 1930s Dutch barge which carried champagne to Paris. After staying in the French capital after the war, it was moved to London just before the millennium. It also has a sister ship, The Battersea Barge.

The Tapestry *1 Lower Richmond Road, SW14 7EZ*
Once called the Jolly Milkman, which has Benny Hill or Pat Mustard connotations depending on your preferred era of comedy, this was rechristened in 2001 to commemorate the tapestry works which once stood in Mortlake in the 17th century. They produced works which can

still be viewed today in museums and country houses across England, and even further afield to the Metropolitan Museum of Art in New York and even a palace in Bratislava. The name also almost fits as a description of this pub restaurant's food offering, Tapas!

Tapping the Admiral *77 Castle Road, NW1 8SU*
Following the Battle of Trafalgar and Lord Horatio Nelson's demise, his crew famously decided to preserve his body in a barrel of brandy to ensure he had a full funeral back in the UK. On the way, they decided to raise a toast to his victory by drilling a small hole in the barrel and, using a macaroni straw, they would tap the barrel to take a sip from the brandy which gave rise to the saying 'Tapping the Admiral'.

Tarmon Free House *243 Caledonian Road, N1 1ED*
This pub on the Caledonian Road is so named because the original landlord was from Tarmonbarry in Ireland.

Tattersall's Tavern *2 Knightsbridge Green, SW1X 7QA*
Named after a horse auctioneer's called Tattersall's who had their sale rooms on Knightsbridge Green between 1865 and 1939.

Telephone Exchange *10-18 London Bridge Street, SE1 9SG*
A wartime telephone exchange was set up in this building in 1940 and it was this legacy Fullers sought to tap into when they opened the pub in 2019, replacing a bar which had previously traded here.

Temperance *90 Fulham High Road, SW6 3LF*
A nod to the building's original use as a billiard hall run by Temperance Billiard Hall Company and part of the wider temperance movement at the time. We wouldn't be the first people to point out the irony that it now houses a cavernous pub!

The Telephone Exchange.

Theodore Bullfrog.

Ten Bells *84 Commercial Street, E1 6LY*
Originally merely the Eight Bells on account of the standard complement in situ at Hawksmoor's Christ Church. When the church upped the ante and got themselves ten of the best in 1788, the pub also rang in the changes and altered their name as a result.

Terminal 6 Lounge *Osterley Park Hotel, 764 Great West Road, TW7 5NA*
Located on the ground floor of the Osterley Park Hotel; coincidentally this is six miles away from Heathrow. At one stage the Airport had a Terminal 6 planned as part of their expansion proposals; however, this was dropped in 2017 as part of cost cutting measures, even before the whole new runway plans ran into trouble in 2020.

Theodore Bullfrog *26-30 John Adam Street, Charing Cross, WC2N 6HL*
Close to Charing Cross station, this pub plagued us for months, and still does. The pub sign is deceptive in that it shows a very smartly dressed bullfrog. The only tenuous link between Theodore and Bullfrog is a 1990s Canadian children's TV show called *Theodore Tugboat* (imagine *Thomas the Tank Engine* but with boats). In one episode there was a bullfrog who scares the boats. The pub was renamed in 1999, so this is a possible albeit slightly far-fetched story. We even tracked down people involved with Regent Inns, the company which first opened the pub as the Theodore Bullfrog, and even they didn't know anything beyond the fact they thought a woman in their marketing department came up with the name. If anyone can shed any more light on this, please get in touch!

The Thomas Cubitt *44 Elizabeth Street, SW1W 9PA*
Thomas Cubitt was a master builder whose mark on London is not to be understated. Responsible for neighbourhoods such as Bloomsbury, Belgravia and Pimlico, as well as other schemes like the east face of Buckingham Palace. Outside of London his projects included Kemp Town in Brighton. His brother Lewis also famously designed King's Cross station. The pub which pays homage to him can be found in Belgravia.

Thomas Neale *39 Watney Market, E1 2PP*
Named after the man who first developed the Shadwell area in the 17th century, Neale was also responsible for the development of the Seven Dials area near Covent Garden, including the alley Neal's Yard which has since spawned the Neal's Yard Remedy business.

Thornbury Castle *29A Enford Street, W1H 1DN*
There is a fair amount of mystery as to why a pub in Marylebone came to be named after a castle in Gloucestershire. The current landlady's theory is that as Henry VIII had acquired both the area of Marylebone and also Thornbury Castle during his time on the throne, the Victorians who first opened this pub in 1852 sought to mark that link.

Three Cranes *28 Garlick Hill, EC4V 2BA*
A nod to the three timber cranes which once stood in nearby Vintry Wharf and, as you may have already guessed, were used to load up wine brought in on the Thames by boat.

Three Greyhounds *25 Greek Street, Soho W1D 5DD*
This pub picks up its name from back when Soho was a hunting ground (which is hard to imagine now!) Greyhounds would have been used to hunt hares back in the day and this pub remembers then.

The Three Johns *73 White Lion Street, N1 9PF*
The three Johns in question are a collection of 18th century radicals, John Glynn, John Wilkes and John Horne Tooke: Islington's link with radical thinking again coming to the fore! In the late noughties it endured rapid changes of identity, being briefly a Hobgoblin and then the Fallen Angel before reverting to its original title. The pub is also famous for hosting a pivotal meeting of the 1903 London Conference of Russian Communists, where Lenin and his allies won a key vote for control of the party's journal. Securing control of the journal led Lenin to call his group 'Bolsheviks' (meaning majority in Russian) and within 14 years they'd seized power of the world's largest country and changed the course of 20th century history forever…

Thornbury Castle.

Three Greyhounds.

Three Lords *27 Minories, EC3N 1DE*
Named after a trio of conspirators, the Earl of Kilmarnock, Lord Lowe and Lord Balmerina, involved in plotting on behalf of Bonnie Prince Charlie during the Jacobite rebellion, and who were all executed at Tower Hill just moments away from where the pub stands today.

Three Steps *High Street, Cowley, Uxbridge, UB8 2DX*
Named after the three steps in the nearby rivers which flow into Little Britain lake. The pub also has three steps which link each of its sections inside.

The Tichenham Inn *11 Swakeleys Road, Ickenham, UB10 8DF*
Tichenham is how Ickenham was recorded in the Doomsday Book of 1086 which was put together at the request of William the Conqueror to document all of the settlements in the country at the time.

Tide End Cottage *8 Ferry Road, Teddington, TW11 9NN*
Teddington Weir, where this pub is located, marks the end of the tidal section of the Thames which then runs eastwards uninterrupted until it enters the North Sea.

Tigers Head *Watt Lane, BR7 5PJ*
Takes its name from the tiger's head on the family crest of the Walsingham's, a noble family who lived nearby at Scadbury Hall during the Tudor and Stewart era and owned an inn on the site where this pub stands today. Sir Francis Walsingham was a major figure during Elizabeth I's reign, serving as her principal secretary for 17 years and considered to be her 'spymaster'.

The Tipperary *66 Fleet Street, Temple, EC4Y 1HT*
Of course, Tipperary is the famous town in the middle of Ireland where the war song "It's A Long Way to Tipperary" comes from. This pub on Fleet Street claims to be the first Irish pub outside the island of Ireland and that it was the first pub to serve Guinness. How it got its name is a bit more complicated as the plaque outside the pub claims it was originally called the Boar's Head and taken over by Mooney's Brewery in 1700 with it being renamed the Tipperary just after the First World War. However blogger Martyn Cornell, writing as Zythophile, has dug deep into this and found this is mostly "nonsense" in his words. Mooney's brewery did take over the pub but not until the 1800s and the pub was renamed after Greene King took over in the 1960s. The same applies to its claim that it was the first Irish pub in London, which he also believes to be nonsense!

The Toll Gate *26-30 Turnpike Lane, N8 0PS*
Prior to the advent of the railways, the fastest way to travel was along the toll roads. Pikes were used as gates to stop the coaches from proceeding until they paid their fare. Once the railways took off, the

traffic declined on the roads and in the 1870s all tolls were abolished. The Tokenhouse.
This Wetherspoons remembers this piece of history near Turnpike
Lane. Although there is the M6 Toll motorway in the Midlands.   Tom Cribb.

Tokenhouse *4 Moorgate, EC2R 6DA*
Named after a nearby building from the 17th century which distributed
legal tender in exchange for tokens, once the individual had amassed
enough, that were used by traders as copper coins rarely issued by the
Government at the time. This custom is also recalled by Tokenhouse
Yard, a minute's walk from the pub.

The Tommy Flowers *50 Aberfeldy Street, Aberfeldy Village, E14 0NU*
Tommy Flowers was an engineer and a pioneer of computer science.
During the Second World War he worked with Alan Turing and was
responsible for designing the Colossus, which is considered the first
programmable computer. Flowers was born a short distance away from
the pub that now bears his name.

Tommy Tucker *22 Waterford Road, SW6 2DR*
This Fulham gastropub looked back to Cockney rhyming slang when it
opened in 2015, its name being the slang for supper. Tommy Tucker is
also the subject of an old rhyme from the Georgian era about a boy who
had to sing for his supper.

The Tommyfield *185 Kennington Lane, SE11 4EZ*
We're off to Oldham for the name of this pub. Tommyfield Market has been running in Oldham for centuries and it acquired its name from the landowner, Thomas Whittaker. The market also claims to have had the first fish and chip shop in the country.

Tom Cribb *36 Panton Street, SW1Y 4EA*
Named after its landlord in the 1820s and early 1830s, Cribb was a legendary champion bare-knuckle boxer and so definitely not someone to pick a fight with over the price of a pint...

Tooke Arms *165 Westferry Road, E14 8NH*
Named after Reverend William Tooke, the man who owned the land on which the original pub here was built. You'll see we said original: this is because the first Tooke Arms was demolished, alongside the Tooke Street it sat on, as part of a 1960s redevelopment of this part of the Isle of Dogs, and replaced with the Barkantine Estate. A new pub with the same name was provided as part of the development; however, Tooke Street itself vanished off the map forever.

The Torch *1-5 Bridge Road, Wembley Park, Wembley, HA9 9AB*
Found near Wembley Stadium, this pub is named after the Olympic Torch which was used in the 1948 London Olympics.

Toucan *19 Carlisle Street, Soho, W1D 3BY*
This Irish pub in the heart of Soho is named after the colourful bird with the long bill which was used in Guinness adverts from 1935 up until the 1980s. Its era includes design classics like 'Guinness is good for you', and there are plenty of Toucans beaming at you within the pub, too.

Town of Ramsgate *62 Wapping High Street, E1W 2PN*
This riverside pub claims to be one of the oldest in the capital. It was originally named the Red Cow before changing to its present name after the home of the fishermen who used to moor here whilst waiting to sell their goods at Billingsgate Fish Market. The Old Stairs are just outside and prisoners were often kept in the cellar before being shipped off to Australia.

Traders Inn *52 Church Street, NW8 8EP*
This name reflects the fact that when the pub first opened, Church Street was home to a thriving market which dealt in hay, vegetables and general goods. The original market site was damaged during the blitz and subsequently redeveloped as part of the Church Street Estate. A market returned to the street in the 1960s, selling antiques and nick-nacks as well as more typical market produce, this was then expanded indoors in the next decade when an old department store was converted into an antique market.

Trafalgar Tavern *Park Row, SE10 9NW*
Now we know there are countless pubs across the UK called the Trafalgar, but we feel the reason behind this one getting the name is extra special and so worthy of the cut. It opened in 1837 and gained its name due to the fact Admiral Nelson's coffin had lain in state in Greenwich Hospital (now the old Naval College) after the Battle of Trafalgar in 1805, right beside where the pub was built.

Tram and Social *46-48 Mitcham Road, Tooting, SW17 9NA*
This building was once a tram shed before conversion into a pub in the 1990s. It's slightly hidden away down a courtyard off the main road. It's not the only striking building hidden behind a more functional facade on Mitcham Road: the Gala Bingo Hall in the former Granada cinema is two doors up and its interior really has to be seen to be believed.

Traitors Gate *14 Trinity Square, EC3N 4AA*
Taken from one of the entrances for prisoners into the Tower of London, particularly well-used during the Tudor period.

Traveller's Home *60-66 Long Lane, DA7 5AR*
It is said that the husband-and-wife team who first set up a pub here, Marey and Robert Mann, were licensed hawkers (the door-to-door salesman of their time) and initially lived on the site in a travelling caravan. When they decided they'd had enough travelling around they built four cottages on the land, turning two of them into this pub.

The Trinity Bell.

Trinity Bell *20 Creechurch Lane, EC3A 5AY*
The pub stands next to St Katherine Cree's Church, one of the few buildings in the city which survived the Great Fire of London. The church originally formed part of the medieval Priory of the Holy Trinity, hence the Trinity Bell name.

True Craft *68 West Green Road, Tottenham, N15 5NR*
There are two layers to this one, the first being it's a pub serving a wide range of craft beers, the second that they also do their own homemade pizzas, the act of making them also being a 'true craft'.

Tudor Barn *Well Hall Pleasaunce, Well Hall Road, Eltham, SE9 6SZ*
Plenty of pubs across London and beyond deploy the mock Tudor frontage; however, this place is the real deal, dating from 1525, but actually minus the usual timber and white frontage so favoured by the imitators.

Turner's Old Star *14 Watts Street, E1W 2QG*
The Turner in question here is none other than renowned 19th century British painter JMW Turner who inherited two cottages on this site in the 1820s, converting them into a pub called The Old Star. The pub's folklore continues with rumours that he set up his then mistress, Sophia Booth, as landlord here, although records of landlords here from this era have no reference to a Mrs Booth. Either way, the pub became 'Turner's Old Star' in 1987.

Twelve Pins *263 Seven Sisters Road, Finsbury Park, N4 2DE*
The Twelve Pins name sadly has nothing to do with nearby Rowan's bowling alley. It is, in fact, named after a mountain range in Galway, Ireland. They are also called the Twelve Bens.

Two Chairmen *39 Dartmouth Street, Westminster, SW1H 9BP*
One of the oldest pubs in Westminster, the pub sign indicates where the name comes from. It depicts two men carrying a sedan chair in which they would carry home the wealthy after the cockfights which took place at the Royal Cockpit theatre which used to sit opposite. You could usually find sedan chairs for hire where blue posts were (see also Blue Posts).

Two Towers *201 Gipsy Road, SE27 9QY*
As well as the glass palace which lent its name to the wider area, Crystal Palace Park was also home to two tall, thin water towers, designed by no less than Isambard Kingdom Brunel. These brick towers survived the fire that destroyed the Palace in 1936 but were dismantled and demolished between autumn 1940 and spring 1941 at the height of the blitz, as they were thought to be an unhelpfully useful landmark for the Luftwaffe.

Vaulty Towers *34 Lower Marsh, SE1 7RG*
This puntastic title references the Vault Theatre, which is just opposite the pub, in the unique setting of the vaults under Waterloo Station. Interestingly, this was not one of the numerous misspellings of the *Fawlty Towers* name in the opening credits of the classic BBC comedy.

Viaduct Tavern *126 Newgate Street, EC1A 7AA*
Named after the nearby Holborn Viaduct which was constructed in the 1860s.

Victoria Stakes *1 Muswell Hill, N10 3TH*
Located at the opposite end of Alexandra Park to the Starting Gate, this pub name also commemorates the fact there was once a racetrack in the park, and the building here originally began life as a coach house and stables. The Victoria Stakes is a horse race that has taken place in Toronto since 1803.

Virgin Queen *94 Goldsmiths Row, E2 8QY*
It was the mock Tudor frontage of this pub which led the present owners to name it after the family's last monarch. It acquired its present title in 2015; prior to that it had been called Albion in Goldsmiths Row, the Albion in question being West Bromwich, probably one of the last things you'd be expecting to find in Hackney. The owner had even got permission from the club to use the team's crest on the pub's exterior.

The Volunteer *245-247 Baker Street, NW1 6XE*
This name harks back to the Volunteer Regiments which were formed during the Napoleonic Wars. Local taverns would display signs calling for volunteers to sign up to the army and defend the country. This tradition continued throughout the nineteenth century and into the twentieth when they were again used during the First World War.

The Viaduct Tavern.

The Virgin Queen.

The W B Yeats *20 Fonthill Road, Finsbury Park, N4 3HU*
W B Yeats was a famous poet and writer at the start of the 20th century. In 1923 he was awarded the Nobel Prize for Literature and was a supporter of Irish Nationalism.

Walker Briggs *23 Westow Hill, Norwood, SE19 1TQ*
This pub is named after two local shops that used to be in the area.

The Walrus & Carpenter *45 Monument Street, EC3R 8BU*
Another literary connection to this pub name. 'The Walrus & Carpenter' takes its name from the poem that is read by Tweedledum and Tweedledee to Alice in *Through the Looking Glass* by Lewis Carroll.

The Watch House *Lewisham High Street, SE13 6JP*
The pub stands on the site of Watch House Green, a village green in Lewisham where the activities varied from criminals in the stocks through to preachers out spreading the word of God. Pub life probably seems sedate in comparison.

The Watchman *184 High Street, New Malden, KT3 4ES*
This pub is based in a former police station built in 1890 and recalls the individuals who used to keep order before the formation of the Metropolitan Police in 1825. Prior to this, local watchmen would be hired to watch over apprehended criminals before they were taken to the magistrates when they opened the following morning. The name was chosen as the local police station before this one, located in Kingston, was known as the 'Watch House', as it was the meeting place for the borough's watchmen.

The Water Rats *328 Gray's Inn Road, WC1X 8BZ*
Named after the pub's owners, the Grand Order of Water Rats, a charitable organisation for people in the entertainment industry. They acquired the pub in 1986 and renamed it in 1992, it having previously been known by the equally unique name of Pindar of Wakefield. Perhaps unsurprisingly, given their trade, they have also continued its status as a live music venue. Pre-Water Rats era names such as Bob Dylan and The Pogues played here; in its present guise it hosted early gigs for Oasis and even Katy Perry…

Watson's General Telegraph *108 Forest Hill Road, SE22 0RS*
Before the electric telegraph, Barnard Watson invented the fastest way of communicating by using semaphore signals which were hoisted on towers which would signal information from ports. He had four lines across the country, including one that ran from Kent to London. Unfortunately, the dense pollution in Victorian London was enough to block the flags. He was declared bankrupt in 1842. The pub is located close to where the final station of his line from Kent was sited.

Waxy O'Conners *14-16 Rupert Street, W1D 6DD*
Waxy O'Connor was born in Dublin in 1788 and, as you can probably guess, he was a candlemaker which is why Waxy is probably a nickname. His capacity for drink was legendary and a tree was planted in his honour. When it died naturally in the 1990s, bits of it were transported to England and are now located in the pub. There is also a sister pub called Waxy's Little Sister round the corner.

We-Anchor-in-Hope
320 Bellegrove Road, Welling, DA16 3RW
The Anchor coming before the hope in this name puzzles the eye at first, a bit like seeing a shop called Spencer and Marks. It refers to the crew of the Mayflower who, as we know from that particular Thameside pub, set off for the 'new world' back in 1620.

The Weir *24 Market Place, Brentford,TW8 8EQ*
The building here reportedly dates back to the early 17th century and began life as The White Horse and was once the residence of the English landscape painter JMW Turner, evidently before he set off east for Wapping. It was renamed The Weir in 2002. The name comes from the fact it backs onto a weir on the river Brent, which also serves as the southernmost point of the Grand Union Canal. If you were to traverse the full length of the canal in the opposite direction, you'd be in Birmingham!

Well and Bucket *143 Bethnal Green Road, E2 7DG*
This is thought to originate from the fact there was a well by the site of the pub, which potentially even supplied the water to help make the beer here. This name vanished off the London pub scene for an extended period; it had a spell in the 1980s as Stick of Rock, a music pub managed by Steve Bruce (not the football manager/crime novelist) before then laying derelict until revived in 2013.

The White Cross *Riverside House, Water Lane, Richmond, TW9 1TH*
This name marks the fact that the pub stands on the site of the Convent of Observant Friars which was built at the end of the 15th century; the order's insignia was a white cross.

The White Hart *Various - Eighteen Pubs in London*
Another hugely popular pub name here. A hart is another name for a male deer. During the 14th century, the White Hart became the emblem of King Richard II. He demanded it be displayed by taverns after many started displaying a Red Lion in support of his uncle, John of Gaunt, following his suppression of the Peasants Revolt in 1381 when he betrayed the protesters and killed their leadership, including one of their leaders, Jack Straw.

White Cross.

White Horse & Bower.

White Horse and Bower *86 Horseferry Road, SW1P 2EE*
Another horse related name on Horseferry Road, this one was chosen as the pub as it was once a watering stop for horses. A bower is a name for a shelter that would have also existed here for our equine friends.

White Lion *Various - Six Pubs in London*
The White Lion was the emblem of King Edward IV. He was one of the central figures in the War of the Roses and was King twice as his house of York battled the House of Lancaster.

The Whittington Stone *53 Highgate Hill, Archway, N19 5NE*
Just outside the pub is a small stone monument, allegedly the location where the fictional Dick Whittington heard the Bow Bells (See The Bow Bells) after giving up on London life. Upon hearing the bells, he turned back into the City and made his success. The real Dick Whittington who the pantomime character is based on was Lord Mayor of London four times.

Who'd a Thought it? *7 Timbercroft Lane, SE18 2SB*
One of the five idlers of Plumstead Common, the four other pub names were all accompanied by a postscript as to how they didn't deliver on their usual function, so there was the Star (which doesn't shine in the sky), the Woodman (who doesn't cut down trees), the Ship (which cannot sail the seas) and the Mill (which doesn't grind corn). You'd look at those four and their respective failings and think: well, 'Who'd a Thought it?'

Wibbas Down Inn *6-12 Gladstone Road, SW19 1QT*
Wibbas Down was an early spelling of what would eventually become
the Wimbledon we all know today.

The Who'd 'a'
Thought It.

The Widow's Son *75 Devons Road, Mile End, E3 3PJ*
A sad tale behind the name of this East End pub. It's found on the
site of an old cottage where a widow patiently waited for her son to
return from sea at Easter. On Good Friday, she baked him a hot cross
bun; however, he never returned as he was lost at sea. She continued
to bake a hot cross bun each year in the slender hope he would return.
After she passed and the workmen came to demolish the cottage, they
found her collection of buns hanging up (which was believed to bring
luck). The new landlord, paying respect to her, kept the buns hanging
in the pub and named the pub in his honour. Each year, he invited the
Royal Navy to hang up a new bun at Easter, a tradition that continues
to this day.

Williamsons Tavern *1 Groveland Court, EC4M 9EH*
Once a residence of the Lord Mayor of London, Robert Williamson
bought the building in 1739, turning it into a hotel and tavern. The
Lord Mayor continued to reside in a section of the building up until
1752 when he moved to Mansion House, which remains to this day the
official residence of that position.

William Bourne *Moor Lane, Chessington, KT9 2BQ*
There is a bit of a story here behind this Bourne's identity; it said that a
pub was first set up here by a Mr Bourne and called The Gate. It quickly
became known as Bonesgate in lieu of the tall tales he used to tell about

Williamson's Tavern.

this spot also being a drop-off point for the bodies of plague victims and also a nice play on his name, too. Apparently, the intrigue this caused brought him more custom, as opposed to putting people off! In 2011 the pub was given a refresh and called William Bourne. The name William is not linked to any specific figure, instead being picked as it was felt to be a suitably historic sounding name!

William Camden *Avenue Road, Bexleyheath, DA7 4EQ*
Mr Camden was a late Tudor/early Stewart-era historian and antiquarian who hailed from Chislehurst. He was the first man to write a historical account of the Tudor period in British history, with the first part being published in 1615 and the second volume in 1625, two years after Camden's passing.

William Morris *Various - Three Pubs in London*
London has several pubs that are named after the influential designer and socialist. His works and patterns which had a big focus on nature became a key part of the Arts & Craft movement of the late 19th century. He was also an accomplished writer, credited with coming up with the modern fantasy genre.

William Webb Ellis *21 London Road, Twickenham, TW1 3RR*
Twickenham is synonymous with rugby, given the presence of the England national stadium and this pub has been named after the man who is alleged to have been the founder of the game of rugby, but in recent years it is increasingly considered to be a myth.

The Windjammer *25 Admiralty Avenue, E16 2PN*
This new pub in the Royal Docks gets its name from a type of merchant sailing ship that existed during the transition to steam ships, which makes sense given the location of the pub.

The Windmill *Various - Twelve Pubs in London*
Numerous pubs in London are named this, most of them because they are on the sites of former windmills which would have existed before London was built up.

Winning Post *Chertsey Road, TW2 6LS*
This references the fact Kempton Park racecourse is located a few miles from the pub.

The Wonder *Batley Road, Enfield, EN2 0JG*
Named after a nineteenth century stagecoach which travelled between London and Shrewsbury in just under sixteen hours. A very impressive time in those days.

The Woodbine *215 Blackstock Road, N5 2LL*
Woodbine is a term used for several climbing plants, such as honeysuckles and ivys.

Wood House *39 Sydenham Hill, SE26 6RS*
A reminder of the fact this area used to be largely covered in woodland, something that was already significantly pared back by the time this pub first opened in 1857.

Woodin Shades *212 Bishopsgate, EC2M 4PT*
Takes its name from its first owner, William Woodin, the shades being a reference to his other business dealings as a wine merchant, because, as you may recall from the Old Shades in Whitehall, the word was once a generic term for a cellar or vault.

Woolf and Whistle *Tavistock Hotel, Tavistock Square, WC1H 9EU*
There are two elements at play here: the first is a reference to Virginia Woolf, the early 20th century author and member of the famous Bloomsbury set who ran her publishing company from another building on this very square. The second aspect, whistle, refers back to the fact this pub is based in the Tavistock Hotel, which is marketed as a business hotel and businesspeople wear suits, or a 'whistle and flute' in cockney rhyming slang. Top marks to Tony Smith in the hotel who concocted this name.

Woolwich Equitable *General Gordon Square, Woolwich, SE18 6AB*
This pub is based on the ground floor of the old headquarters of the Woolwich Equitable Building Society, which was based here from 1935 until 1999. The building society had become a bank in 1997 and the next year was bought by Barclays, who retired the brand in 2006.

The World's End.

The Wrong 'Un.

Ye Olde Cheshire
Cheese.

The World's End *Various - Three Pubs in London*
While it makes for a great pub name, we must briefly depart London for Yorkshire to explain this one. The World's End can trace its name to Knaresborough where a lady known as Mother Shipton used to predict events in the early 16th century. She allegedly predicted major events such as the Great Fire of London and the Spanish Armada. She also predicted the end of the world but, sadly, didn't provide a date!

The World's Inn *113-117 South Street, Romford, RM1 1NX*
Romford's Wetherspoons can trace its name from local poet Frances Quarles. In 1635, he wrote a poem titled 'On the World' in which he states: "The World's an Inn, and I her guest"

The Wrong Un *234-236 The Broadway, Bexleyheath, DA6 8AS*
A 'Wrong Un' (also called a Googly) is a type of delivery that a spin bowler can bowl in cricket which is famous for being difficult to play against. The Bexleyheath area has a big connection to cricket with reference to the sport dating back to the 1740s.

Wych Elm *93 Elm Road, KT2 6HT*
Named after the oldest native elm tree species in the UK. These were a common sight in the area around the pub but were sadly lost when Dutch Elm disease ripped through the UK in the 1960s and 1970s, which saw 90% of UK elm trees wiped out.

Ye Olde Cheshire Cheese *145 Fleet Street, EC4A 2BU*
Named after the nation's favourite cheese at the time the pub was opened after the Great Fire. It was also the first new building to be opened after the events of 1666. Nice to know Londoners have always had their priorities correct!

Ye Olde Mitre *1 Ely Court, Ely Place, EC1N 6SJ*
A mitre is a type of headwear worn by religious figures such as bishops. The connection here is that Ely Court, where the pub can be found, used to be the London residence of the Bishop of Ely.

Ye Olde Monken Holt *193 High Street, Hadley, Barnet, EN5 5SU*
The last pub in Barnet before you leave London, the area has a strong history with monks. Meanwhile, a holt is a small wood, primarily for animals. The Battle of Barnet in 1471 took place just to the north of where the pub stands.

Ye Olde Swiss Cottage *98 Finchley Road, NW3 5EL*
The Swiss Cottage was built in the early 1800s on the site of an old cottage used by the tollgate keeper. An oddity in London to see a Swiss chalet in the middle of the busy Finchley Road, the area around the pub and the London Underground station are now named after it.

Yucatan *121 Stoke Newington Road, Hackney, N16 8BT*
When you think of Dalston, we'd wager a guess Mexico isn't the first thing that springs to mind. The reason behind the name is that the landlord's brother owns a bar in this Mexican city; sadly, we don't think it's called Dalston Kingsland.

Zetland Arms *2 Bute Street, South Kensington, SW7 3EX*
Zetland is an old name for the Shetland Islands in the north of Scotland. The name has Norse origins which, as the language developed, became Zetland and the island's postcode is still ZE. Charlie Chaplin's older brother is claimed to have been the landlord in the 1880s.

Ye Olde Swiss Cottage.

Zetland Arms.

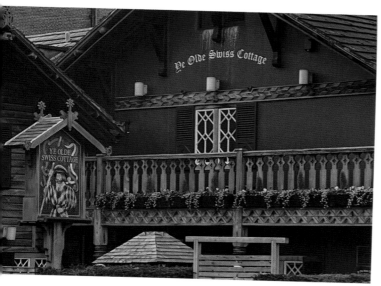

Bibliography

General pub books:

CAMRA's Good Beer Guide 2020
The Dictionary of Pub Names - Walworth Reference Series, 2006
The Old Dog and Duck: The Secret Meaning of Pub Names by Albert
  Jack, 2011

London-wide books:

Derelict London, Paul Talling, Random House Books, 2008 and 2018
Historic Pubs of London, Ted Bruning, Prion Book Ltd, 2000
Inn and Around London: History of Young's Pubs, Helen Osborn,
  1991
London's Best Pubs by Peter Haydon, 2016
Londonist Drinks, 2019
London's Lost Music Venues, Paul Talling, Damaged Goods Books,
  2020
What's In A Name by Cyril M. Harris
What's in a Street Name by Antony Badsey-Ellis, 2017
I Never Knew That About London by Christopher Winn, 2007

Area specific books:

City of London Pubs, Johnny Homer, Amberley Publishing, 2016
Clerkenwell and Islington Pubs, Johnny Homer, Amberley Publishing
  2017
East End Pubs, Johnny Homer, Amberley Publishing, 2018
Hammersmith and Fulham Pubs, Chris Amies, History Press, 2004
Inns, Taverns and Pubs of the London Borough of Sutton: Their
  History and Architecture, A.J Crowe, London Borough of Sutton,
  1980
Southwark Pubs, Johnny Homer, Amberley Publishing, 2017
Pubs of Wimbledon Town (Past and Present), Clive Whichelow,
  Enigma Publishing, 2021
Pubs of Wimbledon Village (Past and Present), Clive Whichelow,
  Enigma Publishing, 2008

Websites

A London Inheritance: www.alondoninheritance.com
CAMRA London Region: www.london.camra.org.uk
Hidden London: www.hidden-london.com
Historic England: www.historicengland.org.uk
J D Wetherspoon: www.jdwetherspoon.com
Layers of London website: www.layersoflondon.org
Londonist Pubs: www.londonist.com/pubs
Nicholsons Pubs: www.nicholsonspubs.co.uk/restaurants
Whatpub.com: www.whatpub.com
Where London's History Happened - In The Pub:
    www.londonspubswherehistoryreallyhappened.wordpress.com
Zyhthophile: www.zythophile.co.uk

What's in a Brand Name?

THE STORIES BEHIND OVER 250 WELL-KNOWN BRANDS

Matthew Wharmby and James Whiting

Have you ever wondered how a famous household brand got its name? The authors of this book have and so they researched the names of over 250 well known brands to satisfy their – and hopefully your – curiosity.

From Asda and Aldi to Zanussi and Zara, we look at names that have been in or on shops either for many years or since only recent times. The only qualification is that they all have something of interest to write about.

Reflecting the on-line age, you will also find Alexa, Cortana and Siri among other names you will not find on the High Street but may at times have been curious about. Plenty of material here for your next pub quiz.

www.capitalhistory.co.uk

ISBN: 978-1-85414-453-9